SAS® Language and Procedures: Introduction

Version 6
First Edition

SAS®

SAS Institute Inc.
SAS Campus Drive
Cary, NC 27513

The correct bibliographic citation for this manual is as follows: SAS Institute Inc., *SAS® Language and Procedures: Introduction, Version 6, First Edition,* Cary, NC: SAS Institute Inc., 1990. 124 pp.

SAS® Language and Procedures: Introduction, Version 6, First Edition

The SAS® System is an integrated system of software providing complete control over data access, management, analysis, and presentation. Base SAS software is the foundation of the SAS System. Products within the SAS System include SAS/ACCESS®, SAS/AF®, SAS/ASSIST®, SAS/CPE®, SAS/DMI®, SAS/ETS®, SAS/FSP®, SAS/GRAPH®, SAS/IML®, SAS/IMS-DL/I®, SAS/OR®, SAS/QC®, SAS/REPLAY-CICS®, SAS/SHARE®, SAS/STAT®, SAS/CALC™, SAS/CONNECT™, SAS/DB2™, SAS/EIS™, SAS/INSIGHT™, SAS/PH-Clinical™, SAS/SQL-DS™, and SAS/TOOLKIT™ software. Other SAS Institute products are SYSTEM 2000® Data Management Software, with basic SYSTEM 2000, CREATE™, Multi-User™, QueX™, Screen Writer™, and CICS interface software; NeoVisuals® software; JMP®, JMP IN®, JMP SERVE®, and JMP Ahead™ software; SAS/RTERM® software; the SAS/C® Compiler, and the SAS/CX® Compiler. MultiVendor Architecture™ and MVA™ are trademarks of SAS Institute Inc. *SAS Communications®, SAS Training®, SAS Views®,* and the SASware Ballot® are published by SAS Institute Inc. All trademarks above are registered trademarks or trademarks of SAS Institute Inc. in the USA and other countries. ® indicates USA registration.

The Institute is a private company devoted to the support and further development of its software and related services.

Other brand and product names are registered trademarks or trademarks of their respective companies.

Doc S19N, Ver 1.25N, 01AUG90

Contents

Reference Aids

Displays

Figures

Tables

Special Topics

Credits

Documentation

Composition | Gail C. Freeman, Cynthia M. Hopkins, Pamela A. Troutman, David S. Tyree

Graphics | Creative Services Department

Proofreading | Gwendolyn T. Colvin, Carey H. Cox, Paramita Ghosh, Beth A. Heiney, Josephine P. Pope, Toni P. Sherrill, John M. West, Susan E. Willard

Technical Review | Deborah S. Blank, Mark E. Britt, Gloria N. Cappy, Brent L. Cohen, Art Cooke, Jodie Gilmore, Barrett Joyner, Brenda C. Kalt, Tasha Kostantacos, Lynn Leone, Lynn N. Patrick, Denise M. Poll, Joy Reel, Lisa M. Ripperton, David C. Schlotzhauer, John Sims, Terry K. Smith, Maggie Underberg, Helen F. Wolfson

Writing and Editing | Amy S. Glass, Christina N. Harvey, Stacy A. Hilliard, Selene Hudson, J. Renee Hurt, Jeffrey Lopes, Kathryn A. Restivo, Judith K. Whatley

Software

Detailed credits for the DATA step, SAS Display Manager System, and other SAS System components are in the front of *SAS Language: Reference, Version 6, First Edition.* Detailed credits for the procedures in the SAS System are in the front of *SAS Procedures Guide, Version 6, Third Edition.*

Acknowledgment

Special thanks to Jane T. Helwig, now a student at the School of Medicine, University of North Carolina, Chapel Hill, NC, who wrote the original *SAS Introductory Guide.*

x

Using This Book

Purpose

SAS Language and Procedures: Introduction, Version 6, First Edition provides basic introductory material about base SAS software, Release 6.06. This book describes base SAS software and explains how to access, analyze, manage, and present data in various formats. As an introductory guide, it is intended to give you an overview of the software; therefore, it does not include detailed descriptions of all of the functions of SAS software. Usage and reference guides for base SAS software provide complete information. These books are described in "Additional Documentation" later in "Using This Book."

"Using This Book" contains important information that will assist you as you use this book. This information includes how much experience is required before using this book, how the book is organized, and what conventions are used in text and sample SAS programs.

Audience

SAS Language and Procedures: Introduction is intended for those of you who want to learn how to use base SAS software and who consider yourselves new users, whether you are a relative newcomer to programming or are an experienced programmer in another language.

Prerequisites

Before you use this book you should be familiar with using your computer terminal keyboard and you should know how to invoke the SAS System on your host operating system. Contact the SAS Software Consultant at your site for information on how to invoke the SAS System. It is also a good idea to have access to other SAS System users for answers to specific questions.

How to Use This Book

This section provides an overview of the information in this book, explains its organization, and describes how you can best use the book.

Organization

This book is divided into 12 chapters that take you step-by-step through commonly used features of base SAS software. Each chapter builds on the previous one; you should read them in order.

Chapter 1, "Getting Started Using the SAS System"
describes the SAS System as an applications system, explains the role of base SAS software, and tells you how to get started using it.

Chapter 2, "Introduction to Programming Using Base SAS Software"
introduces you to base SAS software. It defines and describes SAS programs, SAS statements, SAS data sets, and typical data set components.

Chapter 3, "Creating SAS Data Sets"
explains how to create SAS data sets and describe data to the SAS System.

Chapter 4, "Modifying Data"
shows you how to use SAS statements to create and modify variables, process selected observations, delete or select observations, and create new data sets.

Chapter 5, "Using SAS Procedures"
describes SAS procedures and explains how to use them to analyze, process, and display data. The chapter also shows you how to add and remove titles and footnotes.

Chapter 6, "Rearranging Data Using the SORT Procedure"
defines sorting and explains when and how to sort data with the SORT procedure.

Chapter 7, "Creating Reports Using the PRINT Procedure"
explains how to create and print reports using the PRINT procedure. The chapter also describes how to produce reports containing selected variables, observations, and data groups.

Chapter 8, "Plotting Data Using the PLOT Procedure"
describes how to create plots and control their appearance by changing plotting symbols and specifying tick mark locations.

Chapter 9, "Charting Data Using the CHART Procedure"
explains how to use SAS statements to produce vertical and horizontal bar charts, block charts, and pie charts with the CHART procedure. The chapter also describes how to control the appearance of charts using statement options.

Chapter 10, "Generating Frequency and Crosstabulation Tables Using the FREQ Procedure"
explains how to create frequency and crosstabulation tables using the FREQ procedure. Chapter 10 also describes statement options used to format tables to your specifications.

Chapter 11, "Generating Summary Statistics Using the MEANS Procedure"
describes how to simplify and summarize data with the MEANS procedure. This procedure returns statistical data for numeric variables.

Chapter 12, "Identifying Errors"
explains how the SAS System detects errors and shows you how to prevent common coding errors, understand SAS log messages, and verify that your data are correct.

Appendix 1, "Using the SAS Display Manager System"
>   describes the SAS Display Manager System, an interactive, full-screen facility
>   that you view and operate through a series of windows. The appendix explains
>   how to use display manager to edit text, submit SAS programs, manage
>   program output, and understand the results of your program. Display
>   manager is a convenient way to interact with the SAS System.

What You Should Read

To gain full benefit from *SAS Language and Procedures: Introduction*, you should
read every chapter of this book, beginning with Chapter 1. Chapter 1 provides a
broad overview of the SAS System, explaining how the system works and what
you can do with SAS software. Read Chapter 1 to gain a comprehensive view of
the system. Chapters 2 through 12 are task-oriented. They should be read
consecutively because they build on what you learned in previous chapters. If you
skip chapters, you may not be prepared for the material in later chapters.

Reference Aids

SAS Language and Procedures: Introduction is organized so that you can read it
from cover to cover. However, after you have gone through the book once you
may want to refer to it again. The following sections will help you locate the
information you need:

Contents	lists the chapters with their page numbers. Each chapter includes a table of contents that lists the sections of that chapter and their page numbers.
Reference Aids	lists page numbers for all displays, figures, tables, and special topics in the book.
Glossary	provides definitions of general SAS terms you find in the chapters.
Index	provides the page numbers where specific topics, procedures, statements, and options are discussed.
inside cover graphics	provide functional overviews of the SAS System. The inside front cover depicts the entire SAS System. The inside back cover illustrates how base SAS software is organized.

Conventions

This section explains the various conventions used in presenting text, examples,
and output.

This book often abbreviates the term *the SAS System* to *the system*. This does
not indicate your operating system.

Typographical Conventions

You will see several type styles used in this book. The following list explains the meaning of each style:

roman
: is the standard type style used for most text in this book.

UPPERCASE ROMAN
: is used for SAS statements, variable names, and other SAS language elements when they appear in the text. However, you can enter these elements in your own SAS programs in lowercase, uppercase, or a mixture of the two.

italic
: is used for generic terms that represent values you must supply. This style is also used for special terms defined in the text.

`monospace`
: is used for examples of SAS programs and commands. This style is also used for character values when they appear in text.

In some illustrations, gray shading, red shading, or red highlighting is used to help illustrate specific topics.

Conventions for Examples and Output

Each of the chapters includes examples that illustrate some of the features of base SAS software. Each example contains an explanation of the nature of the example and the sample SAS program. A sample of the output that appears on your display is included with most examples, enclosed in a box. In each chapter, the output is numbered consecutively starting with 1, and each is given a title.

The appearance of output is controlled by the settings of certain SAS system options. SAS system options are instructions that affect the entire SAS session and control the way the SAS System performs operations. System options remain in effect for the entire session unless you respecify them. Most of the programs in this book were run using the following SAS system options:

- □ NODATE
- □ NONEWS
- □ NOMEMRPT
- □ NOSTIMER
- □ LINESIZE=80
- □ PAGESIZE=60.

In examples where other settings are in effect, a note in the text or a footnote explains the different settings.

To change a system option, specify the option in an OPTIONS statement. For example, to print the date automatically on each page of output, use the following statement somewhere near the beginning of your SAS job:

```
options date;
```

All of the examples in this book were produced using one data set named HTWT. You are not expected to work through the examples by entering the data and producing output. However, if you choose to do this, you can. A listing of the data you need to create your own HTWT data set can be found in Chapter 2, Figure 2.1.

Additional Documentation

SAS Institute provides many publications about software products of the SAS System and how to use them on specific host operating systems. For a complete list of SAS publications, you should refer to the current *Publications Catalog*. The catalog is produced twice a year. You can order a free copy of the catalog by writing to

> SAS Institute Inc.
> Book Sales Department
> SAS Campus Drive
> Cary, NC 27513

In addition to *SAS Language and Procedures: Introduction*, you will find these other documents helpful when using base SAS software:

□ *SAS Language: Reference, Version 6, First Edition* (order #A56076) provides detailed reference information about SAS language statements, functions, formats, and informats; the SAS Display Manager System; the SAS Text Editor; and any other element of base SAS software except for procedures.

□ *SAS Procedures Guide, Version 6, Third Edition* (order #A56080) provides detailed reference information about the procedures in base SAS software.

□ *SAS Language and Procedures: Syntax, Version 6, First Edition* (order #A56077) provides complete syntax information for portable base SAS software (it does not include syntax for operating-system-dependent features).

□ *SAS Language and Procedures: Usage, Version 6, First Edition* (order #A56075) provides task-oriented examples of the major features of base SAS software.

□ SAS documentation for your host operating system provides you with reference information for your computing environment.

Chapter **1** Getting Started Using the SAS® System

Introduction

The SAS System is an integrated applications system. Its power, flexibility, and ease of use enable you to gain strategic control of all your data processing needs. To understand the SAS System, you should take a closer look at the features that make it strategic for you and your organization. This chapter describes the SAS System as an applications system, explains the vital role of base SAS software, and tells you how to get started using the SAS System.

Why Is the SAS System an Applications System?

An application is any use you have for your data, and an applications system is software that gives you the tools you need to make the data useful and meaningful. To be strategic (that is, to make your organization most productive), an applications system should give you total control of your data, facilitate applications that run in more than one computing environment, accommodate the skill levels of potential users, and provide a full inventory of applications development tools.

The SAS System is strategic for your applications development needs because it integrates all of these elements into one powerful, flexible, and easy-to-use software system.

Total Control of Your Data

With any body of data, you must perform four basic tasks to make it useful and meaningful. Figure 1.1 shows your data at the center of four basic tasks: access, management, analysis, and presentation. These four data-driven tasks are common to all applications.

Figure 1.1
Four Basic Tasks
Common to All
Applications

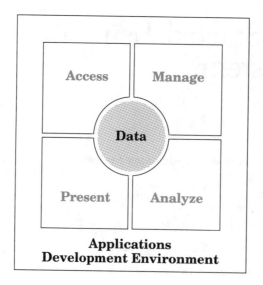

The SAS System enables you to access data for use in your applications, no matter how or where the data are stored or in what format the data exist. You have access to data stored in corporate databases, as well as access to data stored on different computers. Once you access the data through the SAS System, you can use its data management features to update, rearrange, combine, edit, or subset data before analysis. The SAS System's analysis tools range from simple descriptive statistics to more advanced or specialized analyses for econometrics and forecasting, statistical design, computer performance evaluation, and operations research. The SAS System's data presentation capabilities range from simple lists and tables to multidimensional plots to elaborate full-color graphics, both on paper and on your display.

Portable Applications

The SAS System is portable across computing environments. Your computing environment is determined by your hardware and the host operating system running on it. And portability means that your SAS applications function the same, look the same, and produce the same results whether you are using a mainframe, minicomputer, or microcomputer to process your data.

All this is possible because the SAS System has a layered structure called MultiVendor Architecture (MVA). Figure 1.2 shows that MultiVendor Architecture has a host component and a portable component.

Figure 1.2
Layered
MultiVendor
Architecture

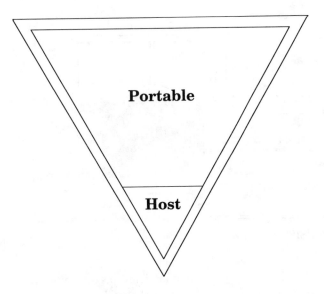

The host component of the SAS System is written separately for each environment. It provides all the required interfaces to the operating system and computer hardware. The host component also allows the SAS System to make use of the technology provided by individual operating environments. For example, you can use the windows, pull-down menus, and icons available in your operating system to work with SAS applications.

Having portable components of the SAS System means not only that your SAS applications run the same in all environments, it also means that you have connectivity. You can develop SAS applications in one environment and run them in other environments without rewriting. And if your site links several types of computers, you can share data and SAS applications across the entire network.

Flexible User Interfaces

The SAS System meets the needs of both the novice user and the most experienced SAS programmer in your organization by providing flexible user interfaces. A user interface is simply a way to create and run SAS applications. Figure 1.3 illustrates the two most common user interfaces.

Figure 1.3
*Two User
Interfaces*

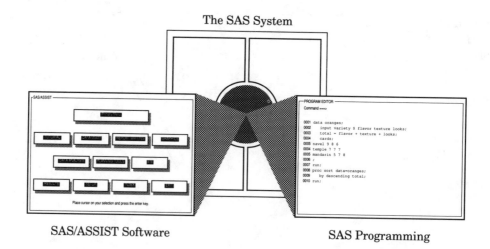

SAS/ASSIST Software   SAS Programming

SAS/ASSIST software is a menu-driven, task-oriented interface. Its menus allow you to select keywords that describe such tasks as managing data, printing reports, or creating graphics. New SAS users can use SAS/ASSIST software to develop applications without learning the syntax of the SAS language because SAS/ASSIST software actually builds and stores SAS programs. More experienced users can use SAS/ASSIST programs as a base for customized applications that run within the SAS/ASSIST environment.

You can also develop SAS applications by writing SAS programs. If you choose to write SAS programs, the SAS Display Manager System is a convenient programming environment. Display manager is an interactive windowing system that lets you write and modify your programs, run them, and monitor the output and messages. Its convenient pull-down menus simplify routine file-management tasks. And you can even customize the display manager work environment by creating your own pull-down menus. Figure 1.4 shows the windows of a typical display manager session.

Figure 1.4
*SAS Display
Manager
Windows*

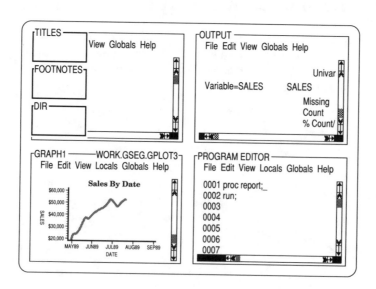

Complete Applications Development Environment

The SAS System is a powerful programming language and a collection of ready-to-use programs called procedures. Combined with other features of the SAS System, the SAS language and its procedures make possible an unlimited variety of applications—from general-purpose data processing to highly specialized analyses in diverse applications areas. Figure 1.5 shows the four basic data-driven tasks surrounded by a complete applications development environment.

Figure 1.5
Applications
Development
Environment

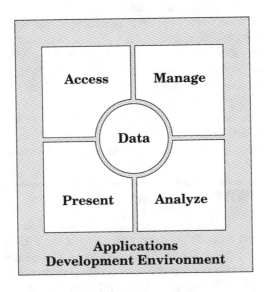

To develop an application, you must write SAS programs and provide a means for running them, often adding some way for the user to interact with the application as it runs. The windowing features of SAS/AF and SAS/FSP software help you build menu-driven, interactive applications for other users.

The ability to create customized, menu-driven applications is enhanced by two language facilities that have their own syntax and programming structures. The SAS macro language, available with base SAS software, enables you to generate and store text strings and communicate information from one program to another. With the Screen Control Language (SCL) of SAS/AF and SAS/FSP software, you can control all aspects of an interactive application, for example building menus, handling errors, controlling program branching, and customizing messages to the user.

In summary, the SAS System is an applications system that gives you total control over your data, provides hardware independence, accommodates the skill levels of your users, and provides a complete applications development environment. Its integrated approach to software development satisfies your diverse data, user, hardware, and applications needs.

What Is the Role of Base SAS Software?

Looking at the graphic on the inside front cover, you see that the SAS System can be represented by a wide range of products grouped within applications areas. Base SAS software is the cornerstone that supports these applications areas. The SAS language, with its statements, functions, formats, and informats, is integral to the SAS System because it forms the building blocks from which all SAS applications are created. Combined with the general-purpose base product procedures, the SAS language gives base SAS software all the functionality required to access, manage, analyze, and present your data.

Equally important is the strategic role base SAS software plays in supporting flexible programming interfaces with display manager, the SAS Text Editor, and the macro language. In addition, base SAS software supports the MultiVendor Architecture that gives the SAS System its hardware independence.

Because the base product is fundamental to the SAS System, this book features simple examples that give you a flavor of the syntax and structure of base SAS software programming. The examples start with some data, get them into a form recognized by the SAS System, and then use base SAS software procedures to create reports. By working through these examples, you become familiar with the SAS System's ability to access, manage, analyze, and present data.

How Do I Start Using the SAS System?

There are two different ways to produce results with the SAS System. One way is to write SAS programs; the other is to use SAS/ASSIST software. This book tutors you in writing simple SAS programs. Because the SAS language is easy to learn and its statements and procedures are powerful, you can produce immediate and impressive results with just a few English-like statements.

Before you try the examples in this book, take the following steps:

□ Decide which method of operation you will use to run your SAS programs.

□ Learn the appropriate command for invoking the SAS System. Different sites have different procedures and policies for invoking the SAS System.

□ Learn where your results will be printed.

Running SAS Programs

Figure 1.6 illustrates the four ways to run SAS programs. The methods differ in the speed with which they run, the amount of computer resources they require, and the interaction you have with the program (that is, the kinds of changes you can make while the program is running). The examples in this book produce the same results, regardless of the way you run the programs. The following list briefly describes these four methods of running the SAS System:

batch mode
    To run a program in batch mode, you prepare a file containing a SAS program including any statements in the operating system's command language that you need. Submit the program to the computer. The computer's

operating system schedules your job for execution and runs it. Your terminal session is free for you to work on something else while the program runs. The results of your SAS program go to a prespecified destination; you can look at them when the program has finished running.

noninteractive mode

In noninteractive mode, you submit the SAS program to the computer. The program runs immediately and occupies your current terminal session. You usually don't see the results of your SAS program until it has finished running.

interactive line mode

In interactive line mode, you enter one line of a SAS program at a time. The SAS System recognizes steps in the program and executes them automatically. You can see the results immediately on your terminal display.

display manager mode

In display manager mode, you interact directly with the SAS System. Display manager mode is a quick and convenient way to write, submit, and view the results of your SAS programs. Because of its windowing capabilities, it is the easiest and most effective method available. See Appendix 1, "Using the SAS Display Manager System," for more information.

Figure 1.6
*Four Methods
of Running
SAS Programs*

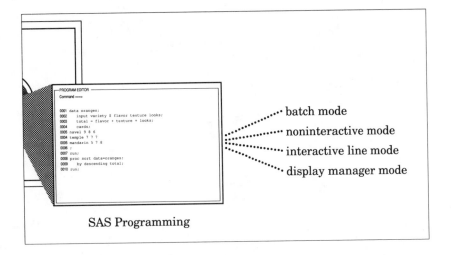

Getting More Help
=================

If you have a problem while working with the SAS System, contact the SAS Software Consultant at your site. If the problem cannot be resolved locally, your SAS Software Consultant should call the Institute's Technical Support Division at (919) 677-8008 weekdays between 9 a.m. and 8 p.m. Eastern Standard Time for assistance with new problems, and between 9 a.m. and 5 p.m. for assistance with existing problems with tracking numbers. Users outside the United States should contact their local SAS Institute office.

Chapter **2** Introduction to Programming Using Base SAS® Software

Introduction

The SAS System is an integrated applications system that provides complete data processing and analysis capabilities. This chapter introduces you to base SAS software, the software product that forms the foundation of the SAS System. Learning to use base SAS software enables you to program with various SAS System features and products.

This chapter also introduces you to the basic components of the SAS System. It defines and describes SAS programs, SAS statements, SAS data sets, and typical data set components. These basic components are discussed in more detail in later chapters.

Processing Data Using the SAS System

The SAS System is a software system composed of computer programs that work together to perform specific tasks. The system reads data, such as letters or numbers, in various forms and organizes them into a SAS data set. A *SAS data set* stores data in a form the system can identify and manage as a unit.

Once the data have been organized into a SAS data set, you can access, analyze, revise, and display the data using one computer program. You do not need to prepare separate programs for different tasks.

Figure 2.1 illustrates a typical SAS data set. The data set contains data that describe participants in an annual health study. The data consist of the name, sex, age, height, and weight of each participant.

Figure 2.1 also illustrates the following data set components: data values, variables, and observations. You should be familiar with these components before you write a SAS program.

Figure 2.1
A SAS Data Set

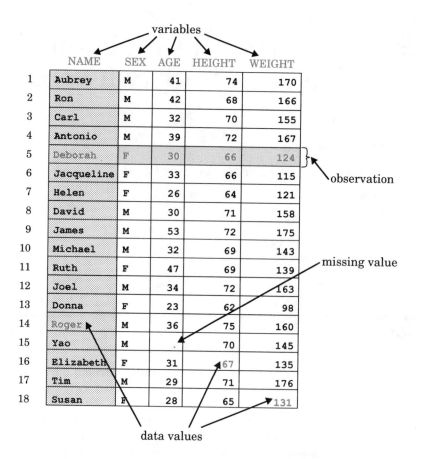

A *data value* is a single unit of information, such as one person's height. Each of the items recorded in Figure 2.1, including Roger's name, Elizabeth's height, and Susan's weight, is a data value.

A *variable* is a set of data values that describe a specific characteristic, for example the heights of all individuals in a group. In Figure 2.1, each column of data values is a variable. The weight values make up the WEIGHT variable, the height values form the HEIGHT variable, and so on.

SAS variables can be classified as character or numeric. *Character variables* contain data values consisting of a combination of letters of the alphabet, numbers, and special characters or symbols. *Numeric variables* contain values consisting only of numbers and related symbols, such as decimal points, plus signs, and minus signs.*

The SAS System identifies variables by name. Variable names can contain from one to eight characters and must begin with a letter or underscore (_). Subsequent characters must be letters, numbers, or underscores. Blanks cannot appear in variable names. When specifying variable names, select descriptive names that reflect the contents of each set of data values. For example, in Figure 2.1 the weight values are named WEIGHT, the name values NAME, the sex values SEX, and so on.

---

* The SAS System also reads *nonstandard data*, or data that appear in other forms, such as numbers containing commas. However, it is beyond the scope of this book to cover this topic. Refer to *SAS Language: Reference, Version 6, First Edition* for additional information.

An *observation* is a set of data values for the same item, for example all physical measurements for one person. Figure 2.1 contains 18 observations. Each observation is a row of information containing the name, sex, age, height, and weight of each person.

A SAS data set contains data values organized into variables and observations. Figure 2.1 contains 18 observations, 5 variables, and 90 data values, one of which is missing.

Missing values represent missing or unavailable data values to the SAS System. Missing values are represented with periods or blanks, depending on the method of data entry and the type of data value. In Figure 2.1, the AGE value for Yao is missing and therefore represented with a single period. Chapter 3, "Creating SAS Data Sets," explains missing values in more detail.

Entering Data into the Computer

Now that you are more familiar with data and how the SAS System stores them, you should know how to enter data in a form the computer can read. Let's assume, for example, that you want to conduct a study analyzing specific physical characteristics, such as height and weight, of a group of people. First, you must obtain the appropriate information for each person, such as his or her name, sex, age, height in inches, and weight in pounds.

Next, enter the data in a form the computer can read. The SAS System allows you to enter data using different methods. For example, you can enter data by putting each variable in specified columns. This method, often referred to as *column input format*, is the most common method of entering data.

In the following example, the names are placed in columns 1 through 10, the sex values in column 12, the ages in columns 14 and 15, the heights in columns 17 and 18, and the weights in columns 20 through 22.

```
----+----1----+----2---
Aubrey     M 41 74 170
Ron        M 42 68 166
Carl       M 32 70 155
Antonio    M 39 72 167
Deborah    F 30 66 124
Jacqueline F 33 66 115
Helen      F 26 64 121
David      M 30 71 158
James      M 53 72 175
Michael    M 32 69 143
Ruth       F 47 69 139
Joel       M 34 72 163
Donna      F 23 62  98
Roger      M 36 75 160
Yao        M .  70 145
Elizabeth  F 31 67 135
Tim        M 29 71 176
Susan      F 28 65 131
```

You also can enter data by separating each value with a space, as illustrated here:

```
Aubrey M 41 74 170
Ron M 42 68 166
Carl M 32 70 155
Antonio M 39 72 167
Deborah F 30 66 124
Jacqueline F 33 66 115
Helen F 26 64 121
David M 30 71 158
James M 53 72 175
Michael M 32 69 143
Ruth F 47 69 139
Joel M 34 72 163
Donna F 23 62 98
Roger M 36 75 160
Yao M . 70 145
Elizabeth F 31 67 135
Tim M 29 71 176
Susan F 28 65 131
```

This method is often referred to as *list input format*. See Chapter 3, "Creating SAS Data Sets," for more information.

Now that you know how to enter data, you are ready to decide what tasks you want the SAS System to perform on your data.

Selecting Tasks

Before you write a program, you must determine what tasks you want the SAS System to perform. For example, let's assume you want the SAS System to print the data after they have been organized into a SAS data set. In addition, you want to produce a graph of the data, with height located on the vertical axis, weight on the horizontal axis, and each point representing the occurrence of an observation.

To produce these results, you must write a SAS program with the appropriate instructions. The following section explains SAS programs in more detail.

SAS Programs

Figure 2.2 illustrates a typical SAS program. A *SAS program* is a group of step-by-step instructions, also known as *SAS statements*, that instruct the computer to perform specific tasks. Figure 2.2 uses shading to illustrate various parts of a SAS program. The SAS statements appear in the shaded areas.

Figure 2.2
A SAS Program

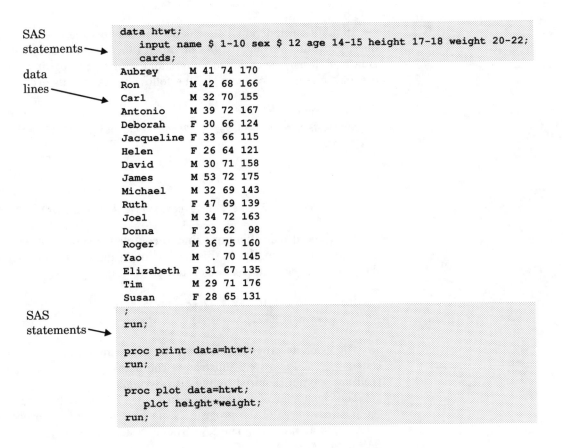

SAS
statements →

data
lines →

SAS
statements →

```
data htwt;
    input name $ 1-10 sex $ 12 age 14-15 height 17-18 weight 20-22;
    cards;
Aubrey     M 41 74 170
Ron        M 42 68 166
Carl       M 32 70 155
Antonio    M 39 72 167
Deborah    F 30 66 124
Jacqueline F 33 66 115
Helen      F 26 64 121
David      M 30 71 158
James      M 53 72 175
Michael    M 32 69 143
Ruth       F 47 69 139
Joel       M 34 72 163
Donna      F 23 62  98
Roger      M 36 75 160
Yao        M  . 70 145
Elizabeth  F 31 67 135
Tim        M 29 71 176
Susan      F 28 65 131
;
run;

proc print data=htwt;
run;

proc plot data=htwt;
    plot height*weight;
run;
```

As you can see, SAS statements usually begin with a *SAS keyword* that identifies the type of statement being used. Common SAS keywords are DATA, INPUT, and PROC. The remainder of the statement contains additional information required for the system to perform the task.

For example, the following DATA statement begins with the keyword DATA, which instructs the SAS System to create a SAS data set. The remainder of the statement, HTWT, specifies the name of the data set you want created.

```
data htwt;
```

Notice that the statement ends with a semicolon (;). All SAS statements end with semicolons.

SAS statements can begin in any column on a line. For clarity, the examples in this book show DATA, PROC, and RUN statements beginning in column 1; all other statements are indented.

In addition, individual statements can occupy one line or extend across several lines, providing variable names or other words are not split between two lines. However, it may be easier to understand your SAS programs when each statement occupies one line. For example, the following statements are valid in either form:

```
data htwt;
   input name $ 1-10 sex $ 12 age 14-15 height 17-18 weight 20-22;
   cards;
```

or

```
data htwt; input name $ 1-10 sex $ 12 age 14-15 height 17-18 weight 20-22; cards;
```

If you need more than one line to complete a statement, extend the statement as needed. The examples in this book contain indented continuation lines to make the programs easier to read, as follows:

```
input name $ 1-10 sex $ 12 age 14-15 height 17-18
      weight 20-22;
```

The example programs in this book also contain blank lines after RUN statements, as Figure 2.2 illustrates. The blank lines make the programs easier to read but are not required by the SAS System.

The remainder of the program in Figure 2.2 consists of data lines, located in the unshaded area of the program. *Data lines* consist of unprocessed data, often referred to as *raw data*. The SAS System reads raw data and organizes them into a SAS data set.

Statements in SAS Programs

Now that you are more familiar with SAS statements, let's take a look at a typical SAS program, as follows:

```
data htwt;
   input name $ 1-10 sex $ 12 age 14-15 height 17-18 weight 20-22;
   cards;
data lines
;
run;

proc print data=htwt;
run;

proc plot data=htwt;
   plot height*weight;
run;
```

Each statement in the program is described as follows:

`data htwt;`
> The first statement is a DATA statement. A DATA statement instructs the SAS System to read data and organize them into a SAS data set. A DATA statement consists of the keyword DATA and the user-supplied data set name. The data set name in this statement is HTWT.

`input name $ 1-10 sex $ 12 age 14-15 height 17-18`
`    weight 20-22;`
> The second statement is an INPUT statement, which provides the information the SAS System requires to organize data into a SAS data set. The INPUT statement begins with the keyword INPUT and contains a user-supplied list of variable names, types, and, if necessary, column locations. In this case, the variable NAME appears first, followed by the variables SEX, AGE, HEIGHT, and WEIGHT.
>
> Notice that the variables NAME and SEX are followed by a dollar sign ($). This symbol indicates that NAME and SEX are character variables with values containing alphabetic characters. The other variables are numeric.

`cards;`
`data lines`
`;`
> The CARDS statement indicates that data lines follow. A single semicolon marks the end of the data lines.

`run;`
> The RUN statement instructs the system to execute the previous statements. Although the SAS System does not always require a RUN statement after the data lines and semicolon, it is recommended that you include a RUN statement in this section of your programs. Once you become more experienced with the SAS System, you can eliminate any unnecessary RUN statements.
>
> Remember that the example programs in this book contain blank lines after RUN statements to make the programs easier to read. The SAS System does not require blank lines after RUN statements. Refer to Chapter 5, "Using SAS Procedures," for more information on RUN statements.

`proc print data=htwt;`
> The PROC PRINT statement instructs the SAS System to print the data. PRINT is a SAS procedure, a prewritten computer program that analyzes and processes data. A PROC statement consists of the keyword PROC, the procedure name, such as PRINT, and, if desired, user-supplied statement options, such as DATA=. The DATA= option specifies the data set name. SAS procedures are described in more detail in later chapters.
>
> **Note:** The SAS System automatically reads the most recently created SAS data set. The DATA= option enables you to override the system default and specify a data set of your choice. The DATA= option also prevents confusion when you are working with two or more data sets. For clarity, the example programs in this book include the DATA= option.

`run;`
> The RUN statement indicates that the previous PROC PRINT statement is ready to be executed.

```
proc plot data=htwt;
```
The PROC PLOT statement requests a plot of the data. The PLOT procedure is described in Chapter 8, "Plotting Data Using the PLOT Procedure."

```
plot height*weight;
```
The PLOT statement provides the details required to produce the plot you want: the variable HEIGHT should be on the vertical axis and the variable WEIGHT on the horizontal axis.

```
run;
```
The RUN statement indicates that the previous statements are ready to be executed.

After writing the program, submit it to the SAS System for processing. (If you don't know how to do this, check with the SAS Software Consultant at your site.) The system then reads your job and carries out the actions you requested, printing the results. These results appear in the next section.

SAS Output

Output 2.1, Output 2.2, and Output 2.3 illustrate the SAS output produced by the following program:

```
data htwt;
   input name $ 1-10 sex $ 12 age 14-15 height 17-18 weight 20-22;
   cards;
data lines
;
run;

proc print data=htwt;
run;

proc plot data=htwt;
   plot height*weight;
run;
```

Output 2.1 is the *SAS log*, which displays the SAS statements you submitted and contains SAS System messages about the execution of your program.

Output 2.1
SAS Log

```
4
5          data htwt;
6             input name $ 1-10 sex $ 12 age 14-15 height 17-18 weight 20-22;
7             cards;

NOTE: The data set WORK.HTWT has 18 observations and 5 variables.

26         run;
27         proc print data=htwt;
28         run;

NOTE: The PROCEDURE PRINT printed page 1.

29         proc plot data=htwt;
30            plot height*weight;
31         run;

32
33

NOTE: The PROCEDURE PLOT printed page 2.
```

The PROC PRINT statement produces Output 2.2. Your data appear here in an organized form, with each column labeled.

Output 2.2
PROC PRINT Output

```
                          The SAS System                            1

        OBS     NAME        SEX    AGE    HEIGHT    WEIGHT

          1     Aubrey       M      41      74        170
          2     Ron          M      42      68        166
          3     Carl         M      32      70        155
          4     Antonio      M      39      72        167
          5     Deborah      F      30      66        124
          6     Jacqueline   F      33      66        115
          7     Helen        F      26      64        121
          8     David        M      30      71        158
          9     James        M      53      72        175
         10     Michael      M      32      69        143
         11     Ruth         F      47      69        139
         12     Joel         M      34      72        163
         13     Donna        F      23      62         98
         14     Roger        M      36      75        160
         15     Yao          M       .      70        145
         16     Elizabeth    F      31      67        135
         17     Tim          M      29      71        176
         18     Susan        F      28      65        131
```

Notice that PROC PRINT automatically displays the number of the observation within the SAS data set in the first column of output.

Output 2.3 contains the plot requested by the PROC PLOT and PLOT statements.

Output 2.3
PROC PLOT
Output

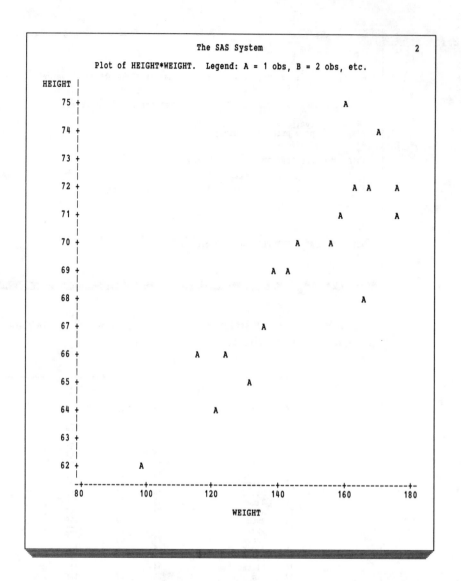

Note that variable HEIGHT is located on the vertical axis, variable WEIGHT is on the horizontal axis, and the height-weight points for the observations are represented by the letter A. The SAS System automatically selects the letter A to represent one occurrence at a point on the output. If two occurrences coincide at a point, the letter B is used; if three coincide, the letter C is used; and so forth.

At the top of each page is the title "The SAS System." The system produces this title automatically unless you specify a title of your own. Refer to Chapter 5, "Using SAS Procedures," for more information on producing titles.

Chapter **3** Creating SAS® Data Sets

Introduction

This chapter focuses on creating SAS data sets. To create a SAS data set, you must produce a program consisting of SAS statements, which instruct the SAS System to perform specific tasks. The following SAS statements are used to create a SAS data set:

□   the DATA statement

□   the INPUT statement

□   the CARDS or INFILE statement.

How the SAS System Creates SAS Data Sets

To understand how to use SAS statements, you first need to know how the SAS System creates a SAS data set. The SAS System, as you learned in Chapter 2, "Introduction to Programming Using Base SAS Software," reads unprocessed or raw data and organizes them into a SAS data set. SAS data sets store data in a form the system can identify and manage as a unit. Once the data are organized in a SAS data set, the system can process the data according to your specifications.
When the SAS System creates a SAS data set, it

□   reads the DATA statement, creates the structure of a SAS data set, and marks the statement as the point to begin processing for each data line

□   uses the description in the INPUT statement to read the data line and produce an observation

□   uses the observation to execute any other SAS statements that are present

□   adds the observation to the data set being created.

It is important to understand that statements are executed once for each data line. Figure 3.1 illustrates this process.

Figure 3.1
*How the
SAS System
Creates Data Sets
from Raw Data*

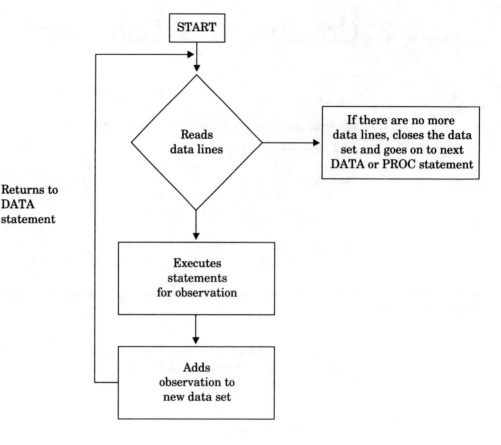

Special Topic: Temporary and Permanent SAS Data Sets

The SAS System creates two types of data sets: temporary data sets and permanent data sets. A *temporary SAS data set* exists only for the duration of the current SAS program or interactive session. Therefore, data stored in temporary SAS data sets cannot be retrieved for use in later SAS sessions; the data set must be re-created each time you begin a new session. (The HTWT data set used in this book is a temporary data set.)

A *permanent SAS data set* exists after the end of the current program or interactive session. Data stored in permanent SAS data sets can be retrieved for use in future programs or sessions.

Both types of SAS data sets have two-level names, of the form

libref.data-set-name

A *libref* is a reference to the name of a SAS data library (a collection of SAS files). A data set is a single file in a SAS data library.

With temporary SAS data sets, the SAS System automatically assigns the libref WORK and you specify the data set name. For example, if you submit the following statement, the SAS System creates a SAS data set named WORK.HTWT:

```
data htwt;
```

The following SAS log message illustrates this:

```
NOTE: The data set WORK.HTWT has 18 observations and
      5 variables.
```

Subsequently, you need use only the data set name when referring to a temporary data set because the SAS System already assumes the default libref WORK. For example, in the following statement, the SAS System automatically prints the data set WORK.HTWT:

```
proc print data=htwt;
```

When you create a permanent SAS data set, you must specify both the libref and the data set name; the SAS System does not assign the libref for you. You must specify a libref other than WORK because the SAS System reserves that libref for temporary SAS data sets. Use a LIBNAME statement, or an appropriate host operating system command, to assign a libref to a SAS data library on your operating system.

When referring to a permanent SAS data set, you typically use both the libref and the data set name. For example, to create a new SAS data set and store it in the permanent SAS data library with the libref STUDY, submit the following statement:

```
data study.htwt;
```

For more information on temporary and permanent SAS data sets, refer to *SAS Language and Procedures: Usage, Version 6, First Edition*; *SAS Language: Reference, Version 6, First Edition*; and the SAS documentation for your operating system.

Now that you are more familiar with SAS data sets, let's look at the statements used to create data sets.

SAS Statements

SAS statements are step-by-step instructions that tell the SAS System to perform specific tasks. Statements also provide other information the system requires to complete these tasks.

You can create a SAS data set using the DATA, INPUT, and CARDS or INFILE statements. These statements are described in the following sections.

Creating SAS Data Sets: the DATA Statement

A SAS DATA statement instructs the SAS System to create and name a SAS data set. DATA statements begin with the SAS keyword DATA and specify the name you select for the data set. For example, the following DATA statement instructs the SAS System to create a SAS data set named HTWT:

```
data htwt;
```

SAS data set names must begin with a letter and contain eight or fewer characters consisting of letters, numbers, or underscores. Remember to select a name that clearly describes the contents of the data set.

Describing Data to the SAS System: the INPUT Statement

The INPUT statement provides the information the SAS System requires to organize data into a SAS data set, such as the variable's name, type, and, if necessary, column location. INPUT statements usually follow the DATA statement, as in the following example:

```
data htwt;
    input name $ 1-10 sex $ 12 age 14-15 height 17-18 weight 20-22;
```

The INPUT statement is important because the SAS System reads the data lines using the description you enter. If the information is not correct, the SAS System produces incorrect results.

The following sections describe two methods used to write INPUT statements and explain data restrictions that determine which method you should use.

Writing INPUT Statements by Specifying Column Location

As you may recall from Chapter 2, you can enter data by placing variables in specified columns. This method, often referred to as column input format, is the most common method of entering and reading data. To write an INPUT statement using this method, complete the following steps:

1. Enter the keyword INPUT:

   ```
   input
   ```

2. Select and enter the name of the first variable:

   ```
   input name
   ```

3. Indicate whether the variable is character or numeric. If the variable's values contain letters or other non-numeric characters, it is a character variable. Character variables are identified by placing a dollar sign after the name of the variable:

    ```
    input name $
    ```

4. Specify the column location for each variable. (As you may recall from Chapter 2, columns 1 through 10 of each line contain the person's name, column 12 contains the person's sex, columns 14 and 15 contain their age, columns 17 and 18 contain their height, and columns 20 through 22 contain their weight.) Write the number of the column where the first data value begins and then enter a dash followed by the column where the data value ends:

    ```
    input name $ 1-10
    ```

    Note: It's important to enter the entire range of columns where the variable's values could be found, even if the values do not occupy all the columns. For example, the NAME value `Donna` occupies just five columns. However, other NAME values, such as `Jacqueline`, occupy all ten columns. When assigning column widths in an INPUT statement, enter an ending column width that accommodates values containing the most characters.

    If the data values for a variable occupy only one column on the data lines, such as the variable SEX here, enter just the column's number:

    ```
    input name $ 1-10 sex $ 12
    ```

    Repeat steps 2 through 4 for each variable that you want the SAS System to read. If your data lines contain values that you don't want read, don't describe these variables in the INPUT statement. The system reads only the information you describe in the INPUT statement. For example, if you need only the NAME and WEIGHT variables, use the following INPUT statement:

    ```
    input name $ 1-10 weight 20-22;
    ```

    After you finish describing the data, complete the statement by entering a semicolon.

    Note: Remember to enter missing data values properly. The SAS System requires a value for each variable; hence, you must represent missing or unavailable character or numeric values with either a single period or blank. Refer to *SAS Language and Procedures: Usage* or *SAS Language: Reference* for more information on missing values.

Writing Simplified INPUT Statements

Another method of entering raw data is by listing variable values without column locations. This method is referred to as list input format. Entering data in list input format is simpler than using the column input method but places specific restrictions on your data. If your raw data meet the following conditions, you can write a simplified INPUT statement to read the data using the list input method:

□ Variable values must contain eight or fewer characters.

□ Each data value on each data line must be separated from the next value by at least one blank column.

□ Missing values, both character and numeric, must be represented by single periods.

□ Numeric values must include any necessary decimal points.

Let's assume you want to write a simplified INPUT statement using the data collected for the height-weight study. If your data satisfy the preceding conditions, complete the following instructions:

1. Begin by entering the keyword INPUT:

    ```
    input
    ```

2. Select a name for the first variable on your data lines. (Remember the naming rules: names must begin with letters and can have no more than eight characters.) Enter the name you choose:

    ```
    input name
    ```

3. Determine if the variable is character or numeric. Character variables contain values consisting of letters or other non-numeric characters. If the variable is a character variable, put a dollar sign after the name of the variable:

    ```
    input name $
    ```

    Repeat the second and third steps until you have listed all the variables. End the INPUT statement with a semicolon, as follows:

    ```
    input name $ sex $ age height weight;
    ```

    When you write your INPUT statement this way, you must list all variables on the input line in succession; it is not possible to skip variables, although you can omit variables at the end of the data lines.

Reading Multiple Data Lines Using INPUT Statements

Because some data lines contain only 80 columns, it often takes more than one line to hold the data for an observation. When you write an INPUT statement that reads multiple data lines for a single observation, you need to let the SAS System know which variables are on which data lines. The pound sign (#) followed by a number tells the system which line contains the next group of variables.

For example, the following statement reads two data lines for each observation. The first line contains the variables WEIGHT and HEIGHT, and the second line contains the variable AGE:

```
input weight 3-5 height 45-46 #2 age 10-11;
```

When you're writing an INPUT statement and have described the variables on the first line, enter a #2 to indicate that the system should move to line 2 before reading any more variables. Then describe the variables on the second line. Continue in this manner if more lines are used for an observation.

Note: Although the SAS System allows you to place a #1 before the first group of variables, it is not necessary to do this because the system automatically begins on the first line.

If you have more than one line per observation but don't need to read any variables from the last line, let the system know how many lines correspond to each observation by entering a # and the number of lines per observation at the end of the INPUT statement. For example, if you have four lines per observation but need to read only from the first two lines, your INPUT statement might look like this:

```
input weight 3-5 height 45-46 #2 age 10-11 #4;
```

Entering Data at the Terminal: the CARDS Statement and Data Lines

The CARDS statement informs the SAS System that data lines immediately follow. CARDS statements usually follow INPUT or INFILE statements in your SAS program, as follows:

```
data htwt;
    input name $ 1-10 sex $ 12 age 14-15 height 17-18 weight 20-22;
    cards;
data lines
;
run;
```

The single semicolon marks the end of the data lines.

Reading Data from Another Computer File: the INFILE Statement

When raw data are stored on disk or tape, you must tell both the SAS System and the computer's operating system where to find the data by using an INFILE statement. INFILE statements begin with the keyword INFILE, which is followed by either the name of the file containing the data or the fileref. *Filerefs*, or file reference names, are the abbreviated names of files, or nicknames.

As the following example illustrates, INFILE statements precede INPUT statements. The following INFILE statement contains the fileref STUDY:

```
data htwt;
    infile study;
    input name $ 1-10 sex $ 12 age 14-15 height 17-18 weight 20-22;
run;
```

The following "Special Topic: Retrieving Data from External Files" describes filerefs in more detail. (If you are not familiar with your operating system's naming conventions, see the SAS Software Consultant at your site for more information.)

Special Topic: Retrieving Data from External Files

The SAS System often needs to read or write raw data to or from an *external file*, a file that is not a SAS data set. To use such a file in a SAS program, you must tell the SAS System where to find it. You can use one of three methods to specify the file location.
You can

□ set up a fileref for the file by using the FILENAME statement and then use the fileref in the INFILE statement

□ identify the file directly in the INFILE statement that uses the file

□ use operating system commands to set up a fileref and then use the fileref in the INFILE statement.

The first two methods are discussed here. The third method depends on the operating system you are using. It is beyond the scope of this book to cover methods for various operating systems. Refer to the SAS documentation for your operating system for additional information.
One method of referencing external files is to use the FILENAME statement to set up a fileref for a file. The fileref functions as a nickname for the external file. You can then use the fileref in later SAS statements that reference the file, such as the INFILE statement.
To reference the raw data stored in a file on your operating system, use the FILENAME statement to specify the name of the file and its fileref. You can then use the INFILE statement with the same fileref to reference the file, as follows:

```
filename study 'your-filename';

data htwt;
   infile study;
   input name $ 1-10 sex $ 12 age 14-15 height 17-18
         weight 20-22;
run;
```

A simple method of referring to an external file is to use the name of the file in the INFILE statement that references the file. For example, if you have raw data stored in a file and you want to read the data, you can tell the SAS System where to find the raw data by putting the name of the file in the INFILE statement. Here is an example:

```
data htwt;
   infile 'your-filename';
   input name $ 1-10 sex $ 12 age 14-15 height 17-18
         weight 20-22;
run;
```

Note that the name of the file must be enclosed in quotes. Other requirements or restrictions may apply, depending on the naming conventions for files on your operating system. For more information, refer to the SAS documentation for your operating system.

Additional information on this topic is available in *SAS Language and Procedures: Usage, SAS Language: Reference,* and the SAS documentation for your operating system.

Chapter **4** Modifying Data

Introduction

This chapter focuses on modifying data. Although raw data can be modified before creating a data set, you may want to use SAS statements to change the data values. This chapter describes the following SAS statements, which are useful when manipulating data:

☐ IF-THEN statement

☐ ELSE statement

☐ DELETE statement

☐ subsetting IF statement

☐ SET statement.

This chapter also explains how to create and modify variables.

Modifying Data Using SAS Statements

Although many SAS programs consist simply of DATA, INPUT, CARDS or INFILE, PROC, and RUN statements, the SAS System provides numerous optional statements that enable you to modify data. For example, some statements create new variables, delete observations, or perform specific tasks based on certain conditions.

As you tackle more challenging data analysis tasks, you may need some of these statements to modify data and tailor your programs to certain specifications. The remaining sections in this chapter describe optional SAS statements used to manipulate data.

Creating and Modifying Variables Using Assignment Statements

The SAS System enables you to create variables and modify existing variable values with assignment statements. *Assignment statements* assign values to specified variables. For example, consider the height-weight study used in earlier chapters. Suppose you need the participants' weights in kilograms rather than in pounds. You can create a new data set and use an assignment statement to modify the WEIGHT variable to obtain the values in kilograms. The following sections explain how to create and modify variables.

Creating Variables

The assignment statement, located after the INPUT or INFILE statement, enables you to create new variables. In the following example, the SAS System creates a new variable containing weights in kilograms by multiplying each weight value in pounds by .45:

```
data htwtkil;
   input name $ 1-10 sex $ 12 age 14-15 height 17-18 weight 20-22;
   wtkilo=weight*.45;
   cards;
data lines
;
run;
```

Notice that this example uses a different data set name to distinguish it from the HTWT data set used in earlier chapters. Also note that the assignment statement begins with the name of the new variable rather than a SAS keyword.

When you create a new variable, the SAS System places another set of data values in your new data set. In addition to the name, sex, age, height, and weight in pounds for each participant in the study, you now also have each person's weight in kilograms, as illustrated in Output 4.1.

Output 4.1
Creating Variables

```
                        The SAS System                              1

     OBS    NAME         SEX   AGE   HEIGHT   WEIGHT   WTKILO

       1    Aubrey        M     41     74       170     76.50
       2    Ron           M     42     68       166     74.70
       3    Carl          M     32     70       155     69.75
       4    Antonio       M     39     72       167     75.15
       5    Deborah       F     30     66       124     55.80
       6    Jacqueline    F     26     66       115     51.75
       7    Helen         F     26     64       121     54.45
       8    David         M     30     71       158     71.10
       9    James         M     53     72       175     78.75
      10    Michael       M     32     69       143     64.35
      11    Ruth          F     47     69       139     62.55
      12    Joel          M     34     72       163     73.35
      13    Donna         F     23     62        98     44.10
      14    Roger         M     36     75       160     72.00
      15    Yao           M      .     70       145     65.25
      16    Elizabeth     F     31     67       135     60.75
      17    Tim           M     29     71       176     79.20
      18    Susan         F     28     65       131     58.95
```

To create a new variable, complete the following step-by-step instructions:

1. Select a name for the new variable. For example, since you used the name WEIGHT to refer to the weight-in-pounds values, you might choose the name WTKILO for the new variable containing the weight-in-kilograms values.

2. Determine which formula is needed to calculate the values of the new variable. To calculate the values of the WTKILO variable, for example, you need to know that one pound is .45 kilograms. Therefore, use the following formula:

   weight in kilograms=weight in pounds x .45 .

3. Write the formula as a SAS statement, putting the new variable name on the left side of the equal sign. For example, the following statement tells the SAS System to multiply the WEIGHT value by .45 (the asterisk means multiply to the SAS System) for each observation in the data set and make the result the WTKILO value for the observation:

   ```
   wtkilo=weight*.45;
   ```

Figure 4.1 illustrates this process.

Figure 4.1
Assigning Variable Values

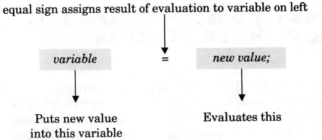

equal sign assigns result of evaluation to variable on left

variable = new value;

Puts new value into this variable

Evaluates this

Modifying Variables

Assignment statements also enable you to modify existing variables in data sets. For example, suppose you want to know each person's height in feet instead of inches. Since you know that one foot contains 12 inches, divide the number of inches by 12, as follows:

height in feet=height in inches divided by 12.

The corresponding SAS statement that follows instructs the SAS System to divide the old HEIGHT value in each observation by 12 (the slash means divide to the SAS System) and to make the result the new HEIGHT value for that observation:

```
height=height/12;
```

This statement seems to indicate that height equals height divided by 12—a confusing impossibility. But the equal sign here informs the system that it should assign the value on the right to the variable on the left, as Figure 4.1 illustrates.

Using arithmetic operators

Table 4.1 lists other symbols, known as *arithmetic operators*, used in assignment statement calculations.

Table 4.1
Arithmetic
Operators

Symbol	Operation	Example	In the SAS System
**	exponentiation	$Y \leftarrow X^2$	`Y=X**2;`
*	multiplication	$A \leftarrow B \times C$	`A=B*C;`
/	division	$G \leftarrow H \div I$	`G=H/I;`
+	addition	$R \leftarrow S+T$	`R=S+T;`
−	subtraction	$U \leftarrow V-X$	`U=V-X;`

Note: The SAS System processes arithmetic operators in the same order of priority as standard mathematical expressions. Items in parentheses, which are used to group parts of mathematical expressions and SAS calculations, are processed before items outside parentheses. In other words, the system processes exponentiation calculations first, then multiplication or division, and then addition or subtraction.

Processing Selected Observations Using IF-THEN/ELSE Statements

Suppose you want to carry out an action for specific observations in a data set. You can use an IF-THEN statement to test observations to determine if certain conditions are true or false. If the condition is true, the SAS System carries out the action specified in the THEN clause, as illustrated in Figure 4.2. If the condition is false, the system continues to the next statement.

Figure 4.2 illustrates the IF-THEN statement structure.

Figure 4.2
How IF-THEN
Statements Operate

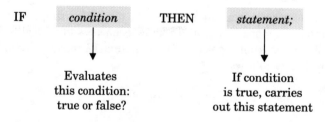

IF   *condition*   THEN   *statement;*

Evaluates   If condition
this condition: is true, carries
true or false?   out this statement

For example, to separate participants into two groups based on age, submit the following statements:

```
data agegrp;
   input name $ 1-10 sex $ 12 age 14-15 height 17-18 weight 20-22;
   if age<40 then group=1;
   if age>=40 then group=2;
   cards;
data lines
;
run;
```

Note that this example uses a different data set name to distinguish it from the HTWT data set that appears in earlier chapters.

As the SAS System processes each observation, it checks the AGE value. If the AGE value is less than 40, the system assigns the variable GROUP in the observation a value of 1. When the AGE value is equal to or greater than 40, the system assigns the variable GROUP a value of 2.

If you create a new variable with an IF-THEN statement, the SAS System creates that variable for all observations. For example, the variable GROUP in the preceding program is created for all observations.

Note: Remember to assign values to all variables or represent missing values accordingly. The SAS System requires values for all variables.

Using comparison operators

Table 4.2 shows the comparison operators you can use in IF conditions. *Comparison operators* are symbols or two-character abbreviations used to form relationships in IF-THEN statements.

Table 4.2
Comparison
Operators

Symbol	Abbreviation	Comparison
<	LT	less than
<=	LE	less than or equal to
>	GT	greater than
>=	GE	greater than or equal to
=	EQ	equal to
¬= or ^= or ~= [1]	NE	not equal to

[1] Use one of these symbols, depending on your terminal.

An IF condition can be a simple comparison of a variable and a value, a comparison of two variables, or a comparison of several variables or values joined by AND and OR operators, as in the following examples:

□   `if age>40 then delete;`

□   `if sex='F' and height gt 70 then size='tall';`

□   `if height lt 65 and weight lt 110 then size='small';`

□   `if name='James' or name='Roger' then delete;`

AND and OR are referred to as *logical operators*. Logical operators, often referred to as Boolean operators, are used in the SAS System to link sequences of comparisons.

The OR operator requires that you enter a variable name each time you enter a corresponding value. For example, the following statement contains the variable NAME twice, once for each value:

```
if name='James' or name='Roger' then delete;
```

Do not omit the second occurrence of the variable NAME; omitting it produces a different expression.

Notice that character variable values, such as `F`, `tall`, `James`, and `Roger`, are enclosed in single quotes. The SAS System requires that you enter such values in quotes. When a value contains an apostrophe, such as the name `O'Brien`, surround the value in double quotes, as follows:

```
if name="O'Brien" then delete;
```

Also remember to use the correct case when specifying values because case may affect the manner in which the SAS System references data. For example, the system will not locate a name with a value of `James` if you assign a value consisting of lowercase letters, as with `name='james'`.

In addition, you must enter the longest character variable value first or enter the appropriate number of blanks before or after the shorter value to make it equal in length to the longer value. For example, the following statement contains value `yes` before value `no`:

```
if age>40 then answer='yes';
if age<40 then answer='no';
```

If value `no` appears before value `yes`, enter one blank before or after value `no`, as follows:

```
if age<40 then answer=' no';
if age>40 then answer='yes';
```

Using ELSE Statements

ELSE statements are used after IF-THEN statements to provide alternate actions when an IF condition is false. If a condition is false, the SAS System omits the THEN action and continues to the next statement. ELSE statements enable you to specify what you want the system to do if a condition is false. For example, the following statements select individuals weighing more than 110 pounds for notification; if the participant's weight is less than or equal to 110 pounds, his or her observation is marked `hold`.

```
if weight gt 110 then action='notify';
else action='hold';
```

Deleting Observations Using DELETE Statements

DELETE statements are useful when you must discard certain observations because you do not need them for your data analysis or because they contain invalid data. Here is an example of a DELETE statement:

```
delete;
```

DELETE statements normally appear as part of IF-THEN statements, as in the following example:

```
if sex='M' then delete;
```

When the SAS System processes a DELETE statement, it checks the IF condition (`if sex='M'`) for each observation. When an observation meets the condition, the system executes the DELETE statement. The system stops working on the current observation, does *not* add it to the data set being created, and returns to the DATA statement.

When the IF condition isn't true for an observation, the system skips the THEN DELETE action, continues executing statements for the observation, and adds the observation to the data set being created before returning to the next observation.

Selecting Observations Using Subsetting IF Statements

Subsetting IF statements select observations with values meeting certain criteria. For example, if you want a data set containing only values for females, you can use a subsetting IF statement to select just those observations where the value of SEX is F. The structure of a subsetting IF statement is identical to an IF-THEN statement without the THEN clause, as follows:

```
if sex='F';
```

This statement informs the SAS System to continue processing the statement for that observation if the value of the variable SEX is F. If the value of the variable SEX is not F, the system stops processing statements for that observation, does not add the observation to the data set being created, and returns to the DATA statement for the next observation.

As the following example illustrates, a subsetting IF statement is exactly the reverse of an IF-THEN statement followed by a DELETE statement. Subsetting IF statements are simply more straightforward and easier to write. The following statements have exactly the same effect with the data used in this book:

```
if sex='F';
```

and

```
if sex ne 'F' then delete;
```

The subsetting IF statement can involve several comparisons or conditions joined by the logical operators AND or OR. For example, if you want to obtain a list of 30-year-old males who are more than 6 feet tall, submit the following statement:

```
if age=30 and sex='M' and height>72;
```

Creating New Data Sets Using SET Statements

Until now, the INPUT statement has described raw data entered at a terminal or data on disk or tape. The SAS System also enables you to create a new SAS data set using observations from an existing SAS data set.

To retrieve data from an existing data set, use the SET statement. Simply enter the word SET in place of the INPUT statement, CARDS statement, and data lines, or in place of the INPUT and INFILE statements. In the following example, the DATA statement prompts the SAS System to create a new data set named HTWT2. The SET statement tells the system to retrieve data from the existing data set HTWT. The HTWT2 data set is an exact copy of the HTWT data set.

```
data htwt2;
   set htwt;
run;
```

If you want to retrieve only selected observations from the original data set rather than duplicate data, use the subsetting IF statement. For example, to create a new SAS data set containing only observations with HEIGHT values under 62, submit the following statements:

```
data height;
   set htwt;
   if height<62;
run;
```

You also can modify data with the subsetting IF statement, as in the following example:

```
data htkil;
   set htwt;
   if height<62;
   wtkilo=weight*.45;
run;
```

Chapter 5 Using SAS® Procedures

Introduction

At this point, you have learned how to use DATA, INFILE, INPUT, and CARDS statements to create SAS data sets. Once you have created a SAS data set, the SAS System can analyze, process, and display data using SAS procedures, or PROCs.

This chapter focuses on SAS procedures. The following sections define SAS DATA and PROC steps and explain how to use SAS procedures to analyze and process data sets. They also explain how to process selected variables, selected observations, and groups of data using the VAR, WHERE, and BY statements. The last section in this chapter explains how to use titles and footnotes in procedure output.

DATA and PROC Steps

All SAS programs consist of a series of statements that, as a group, are designed to accomplish a specific task. These groups of statements are referred to as SAS steps. SAS steps fall into one of two categories: DATA steps and PROC steps.

A SAS *DATA step* is a group of statements that build a SAS data set. SAS DATA steps must begin with a DATA statement and typically contain either INPUT, CARDS or INFILE, or SET statements. DATA steps also can include optional SAS statements used to modify data.

The following example illustrates a typical SAS DATA step:

```
data htwt;
   input name $ 1-10 sex $ 12 age 14-15 height 17-18 weight 20-22;
   cards;
data lines
;
run;
```

Once you have created a data set, the system can analyze, process, and display data using SAS procedures.

SAS procedures are prewritten computer programs that analyze and process data sets. These procedures, or PROCs, read data sets, process data, and display the results. *PROC steps* always begin with a PROC statement, which specifies the name of the SAS procedure you want to run.

To run a SAS procedure, enter the keyword PROC, the name of the procedure, and any optional statement options, followed by a RUN statement. For example, the following statements invoke the PRINT procedure:

```
proc print data=htwt;
run;
```

Combining DATA and PROC Steps

Simple SAS programs, like those used so far in this book, usually consist of a DATA step followed by one or more PROC steps. The following program consists of one DATA step and one PROC step:

```
data htwt;
    input name $ 1-10 sex $ 12 age 14-15 height 17-18 weight 20-22;
    cards;
data lines
;
run;

proc plot data=htwt;
    plot height*weight;
run;
```

However, the SAS System does not require DATA and PROC steps to appear in a specific order. You can write a program that begins with a DATA step, followed by two PROC steps, then a DATA step, and so on, as illustrated in Figure 5.1.

Figure 5.1
SAS Program with DATA and PROC Steps

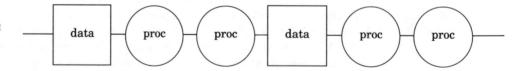

Once you create a SAS data set, the data are available to use at any point in the program. If you use permanent SAS data sets, your programs may not need to begin with a DATA step.

Using PROC Steps

The SAS System enables you to select specific data sets, variables, and observations for processing. The following sections explain how to process selected data sets and data.

Processing Selected SAS Data Sets

If you want a SAS procedure to process a selected SAS data set, simply submit a PROC statement containing the keyword PROC, the procedure name, and the DATA= option followed by the appropriate data set name. For example, the following PROC step prints data in the HTWT data set:

```
proc print data=htwt;
run;
```

Processing Selected Variables

To analyze and process specific variables, use a SAS statement to specify the variable names. For example, the VAR statement enables you to select the variables you want processed and specify the order in which you want them processed.

The following example prints the values of the variables HEIGHT and WEIGHT for all observations in the HTWT data set:

```
proc print data=htwt;
   var height weight;
run;
```

Processing Selected Observations

The SAS System also enables you to process selected observations using the WHERE statement. The WHERE statement selects certain observations from a data set based on a specified condition. Hence, the WHERE statement is similar in action to a subsetting IF statement used in a DATA step.

To define a condition, enter the keyword WHERE followed by the conditions the observations should meet. Combine multiple conditions with SAS operators. For example, the following WHERE statement produces a list of men more than 6 feet tall and 150 pounds:

```
proc print data=htwt;
   where sex='M' and height>72 and weight>150;
run;
```

See Chapter 4, "Modifying Data," for more information on SAS operators.

Processing Data in Groups

The SAS System also enables you to process data in separate, distinct groups using the SORT procedure.* Sorting data arranges observations in order of the values of specified variables.

To sort a data set, enter a PROC statement followed by a BY statement. A BY statement specifies the variables that define the group. For example, the following example sorts the HTWT data set by the variable SEX and creates a new data set, SORTSEX, containing the sorted data:

```
proc sort data=htwt out=sortsex;
   by sex;
run;
```

The OUT= option enables you to retain the unsorted version of the data set for later use. See Chapter 6, "Rearranging Data Using the SORT Procedure," for more information on the OUT= option and the SORT procedure.

You can process a sorted data set with or without a BY statement, depending on the results you need. If you process a sorted data set without the BY statement, the procedure you use processes all observations as one group. For example, the following statements analyze observations in the SORTSEX data set as one group, providing one set of statistics for the data set as a whole:

```
proc means data=sortsex;
run;
```

Output 5.1 shows the results.

Output 5.1
Processing Sorted
Data without
the BY Statement

```
                         The SAS System                              1

         Variable   N      Mean         Std Dev      Minimum      Maximum
         ----------------------------------------------------------------
         AGE       17   34.4705882     7.7630346    23.0000000    53.0000000
         HEIGHT    18   69.0555556     3.5225696    62.0000000    75.0000000
         WEIGHT    18  146.7222222    22.5409576    98.0000000   176.0000000
         ----------------------------------------------------------------
```

However, if you include the BY statement in the PROC MEANS step, as follows, the MEANS procedure recognizes when the value of the BY variable changes and separately processes each group of observations that share the values of the variables in the BY statement. The variables in the BY statement are often referred to as *BY variables*.

```
proc means data=sortsex;
   by sex;
run;
```

---

* Indexing provides a more advanced method of processing data in groups. Refer to *SAS Language and Procedures: Usage, Version 6, First Edition* for more information on this topic.

The resulting Output 5.2 contains two sets of statistics, one for each value of the BY variable SEX.

Output 5.2
Processing Sorted
Data with the
BY Statement

```
                                      The SAS System                              1
---------------------------------------- SEX=F ----------------------------------------

        Variable   N        Mean        Std Dev       Minimum        Maximum
        ----------------------------------------------------------------------
        AGE        7     31.1428571    7.7336617    23.0000000    47.0000000
        HEIGHT     7     65.5714286    2.2253946    62.0000000    69.0000000
        WEIGHT     7    123.2857143   13.8890159    98.0000000   139.0000000
        ----------------------------------------------------------------------

---------------------------------------- SEX=M ----------------------------------------

        Variable   N        Mean        Std Dev       Minimum        Maximum
        ----------------------------------------------------------------------
        AGE       10     36.8000000    7.2541176    29.0000000    53.0000000
        HEIGHT    11     71.2727273    2.0538213    68.0000000    75.0000000
        WEIGHT    11    161.6363636   10.9020432   143.0000000   176.0000000
        ----------------------------------------------------------------------
```

Compare Output 5.1 with Output 5.2.

Adding and Removing Titles and Footnotes

The SAS System enables you to add titles and footnotes to your output with the TITLE and FOOTNOTE statements. These statements are described in the following sections.

Using Titles

The SAS System enables you to enter up to ten titles at the top of your output using the TITLE statement in your procedures. For example, the following statements produce output with titles on the first, third, and fifth lines:

```
proc print data=htwt;
   title 'Height-Weight Study';
   title3 '1990 Statistics';
   title5 'Research Division';
run;
```

Output 5.3 shows the results.

Output 5.3
PROC PRINT with
TITLE Statement

```
                        Height-Weight Study                        1

                          1990 Statistics

                         Research Division

           OBS   NAME        SEX   AGE   HEIGHT   WEIGHT

            1    Aubrey       M     41     74      170
            2    Ron          M     42     68      166
            3    Carl         M     32     70      155
            4    Antonio      M     39     72      167
            5    Deborah      F     30     66      124
            6    Jacqueline   F     33     66      115
            7    Helen        F     26     64      121
            8    David        M     30     71      158
            9    James        M     53     72      175
           10    Michael      M     32     69      143
           11    Ruth         F     47     69      139
           12    Joel         M     34     72      163
           13    Donna        F     23     62       98
           14    Roger        M     36     75      160
           15    Yao          M      .     70      145
           16    Elizabeth    F     31     67      135
           17    Tim          M     29     71      176
           18    Susan        F     28     65      131
```

Note that the SAS System automatically places titles on the first line if you omit the line number in your program. The system interprets TITLE and TITLE1 as equivalent keywords.

To cancel a title for a specific line and all title lines beneath it, enter the keyword TITLE followed by the appropriate line number. This statement is often referred to as a null TITLE statement. The following null TITLE statement cancels titles on the third line and after:

```
proc print data=htwt;
   title3;
run;
```

Output 5.4 shows the results.

Output 5.4
Removing Titles

```
                        Height-Weight Study                        1

           OBS   NAME        SEX   AGE   HEIGHT   WEIGHT

            1    Aubrey       M     41     74      170
            2    Ron          M     42     68      166
            3    Carl         M     32     70      155
            4    Antonio      M     39     72      167
            5    Deborah      F     30     66      124
            6    Jacqueline   F     33     66      115
            7    Helen        F     26     64      121
            8    David        M     30     71      158
            9    James        M     53     72      175
           10    Michael      M     32     69      143
           11    Ruth         F     47     69      139
           12    Joel         M     34     72      163
           13    Donna        F     23     62       98
           14    Roger        M     36     75      160
           15    Yao          M      .     70      145
           16    Elizabeth    F     31     67      135
           17    Tim          M     29     71      176
           18    Susan        F     28     65      131
```

Using Footnotes

The SAS System also enables you to add footnotes to your output with the FOOTNOTE statement. The FOOTNOTE statement functions like a TITLE statement, allowing up to ten footnotes at the bottom of your output. Enter FOOTNOTE statements in your procedure, as in the following example:

```
proc print data=htwt;
   footnote '1990';
   footnote3 'Study Results';
run;
```

Output 5.5 shows the results.

Output 5.5
PROC PRINT with
FOOTNOTE
Statement

```
                             Height-Weight Study                             1

            OBS     NAME         SEX     AGE     HEIGHT     WEIGHT

             1      Aubrey        M       41       74         170
             2      Ron           M       42       68         166
             3      Carl          M       32       70         155
             4      Antonio       M       39       72         167
             5      Deborah       F       30       66         124
             6      Jacqueline    F       33       66         115
             7      Helen         F       26       64         121
             8      David         M       30       71         158
             9      James         M       53       72         175
            10      Michael       M       32       69         143
            11      Ruth          F       47       69         139
            12      Joel          M       34       72         163
            13      Donna         F       23       62          98
            14      Roger         M       36       75         160
            15      Yao           M        .       70         145
            16      Elizabeth     F       31       67         135
            17      Tim           M       29       71         176
            18      Susan         F       28       65         131

                                    1990

                              Study Results
```

As Output 5.5 illustrates, the SAS System processes a footnote without a line number first, as if the keyword FOOTNOTE were followed by the line number 1.

Notice that the output contains the original title generated with previous examples. Titles and footnotes are considered *global statements*, which means that they remain in effect until you cancel them. Global statements can be used in or between DATA and PROC steps.

To cancel a footnote for a specific line and all lines after it, enter a FOOTNOTE statement followed by the appropriate line number. For example, the following statement cancels footnotes on the third line and after:

```
proc print data=htwt;
   footnote3;
run;
```

Output 5.6 shows the results.

Output 5.6
Canceling
FOOTNOTE
Statements

```
                          Height-Weight Study                            1

          OBS    NAME        SEX    AGE    HEIGHT    WEIGHT

           1     Aubrey       M     41       74        170
           2     Ron          M     42       68        166
           3     Carl         M     32       70        155
           4     Antonio      M     39       72        167
           5     Deborah      F     30       66        124
           6     Jacqueline   F     33       66        115
           7     Helen        F     26       64        121
           8     David        M     30       71        158
           9     James        M     53       72        175
          10     Michael      M     32       69        143
          11     Ruth         F     47       69        139
          12     Joel         M     34       72        163
          13     Donna        F     23       62         98
          14     Roger        M     36       75        160
          15     Yao          M      .       70        145
          16     Elizabeth    F     31       67        135
          17     Tim          M     29       71        176
          18     Susan        F     28       65        131

                                1990
```

The SAS System provides an alternate method of entering titles and footnotes with the TITLES and FOOTNOTES windows in the SAS Display Manager System. Refer to *SAS Language and Procedures: Usage* and *SAS Language: Reference, Version 6, First Edition* for more information on these windows.

Chapter 6 Rearranging Data Using the SORT Procedure

Introduction

This chapter explains how to sort data. The following sections define sorting, explain when you need to sort data, and describe how to use the SORT procedure.

Sorting Data

You may often need to analyze and process data in a definite order based on specific criteria. For example, suppose you want to print the observations in the height-weight study alphabetically by student name. Or you might want to analyze data obtained for the females in your study separately from the data for the males. In both cases, you would sort data by one variable: the first sort would be by the variable NAME and the second by the variable SEX.

The SAS System sorts data by rearranging the observations in your data set into an order determined by the values of the variables listed in a BY statement used with the SORT procedure. The following sections explain how to sort data in more detail.

Sorting Data Using PROC SORT

To sort data, enter a PROC SORT statement followed by a BY statement. PROC SORT statements sort SAS data sets by the values of variables listed in BY statements. For example, the following statements sort the HTWT data set by the values of the variable NAME:

```
proc sort data=htwt;
   by name;
run;
```

If the BY statement contains more than one variable, the SAS System sorts the observations in the order the variables are listed. For example, the following statements sort the data by AGE and, within each age group, by NAME:

```
proc sort data=htwt;
   by age name;
run;
```

If two observations contain the same value for the variable AGE, the SAS System uses the value of the variable NAME to determine which observation to list first.

By default, the system rearranges the data set in the order of the variables in the BY statement and writes the rearranged data set, retaining the same data set name. The unsorted version of the data set disappears.*

If you want to keep the unsorted version of the data set, specify the OUT= option in the PROC SORT statement to create another data set containing the sorted version. In the following example, the data set NEW contains the same observations as the data set HTWT, but the observations in the NEW data set are sorted by the values of the variable SEX:

```
proc sort data=htwt out=new;
   by sex;
run;
```

PROC SORT Output

PROC SORT sorts data but does not produce printed output. To view the output, you must invoke another procedure, such as the PRINT procedure. For example, the following program produces and prints the sorted output in Output 6.1:

```
proc sort data=htwt out=sorthtwt;
   by age name;
run;

proc print data=sorthtwt;
run;
```

---

*   Different computer host operating systems may use different sorting sequences. If you run all your SAS programs under a single operating system, you do not need to worry about the sorting sequence. If you move SAS programs from one operating system to another, check with your computing center staff for information on sorting sequences. Refer to the SAS System documentation for your operating system for additional information.

Output 6.1
Data Set Sorted by
AGE and NAME

```
                          The SAS System                          1

        OBS    NAME         SEX    AGE    HEIGHT    WEIGHT

          1    Yao           M       .      70       145
          2    Donna         F      23      62        98
          3    Helen         F      26      64       121
          4    Susan         F      28      65       131
          5    Tim           M      29      71       176
          6    David         M      30      71       158
          7    Deborah       F      30      66       124
          8    Elizabeth     F      31      67       135
          9    Carl          M      32      70       155
         10    Michael       M      32      69       143
         11    Jacqueline    F      33      66       115
         12    Joel          M      34      72       163
         13    Roger         M      36      75       160
         14    Antonio       M      39      72       167
         15    Aubrey        M      41      74       170
         16    Ron           M      42      68       166
         17    Ruth          F      47      69       139
         18    James         M      53      72       175
```

Notice that the SAS System automatically sorts the numeric AGE values in ascending order. It also is possible to request a numeric sort in descending order. Missing values, such as the missing AGE value for Yao, are sorted as the smallest possible values within variable groups. Hence, missing values are listed first within each group.

Also note that the NAME character values are sorted alphabetically within each group. For example, David and Deborah are both 30 years old, but David's name appears before Deborah's. Now compare Output 6.1 with Output 6.2, which was produced before the data were sorted.

Output 6.2
Original Data Set
before Sorting

```
                          The SAS System                          1

        OBS    NAME         SEX    AGE    HEIGHT    WEIGHT

          1    Aubrey        M      41      74       170
          2    Ron           M      42      68       166
          3    Carl          M      32      70       155
          4    Antonio       M      39      72       167
          5    Deborah       F      30      66       124
          6    Jacqueline    F      33      66       115
          7    Helen         F      26      64       121
          8    David         M      30      71       158
          9    James         M      53      72       175
         10    Michael       M      32      69       143
         11    Ruth          F      47      69       139
         12    Joel          M      34      72       163
         13    Donna         F      23      62        98
         14    Roger         M      36      75       160
         15    Yao           M       .      70       145
         16    Elizabeth     F      31      67       135
         17    Tim           M      29      71       176
         18    Susan         F      28      65       131
```

Chapter **7** Creating Reports Using the PRINT Procedure

Introduction

The SAS System supports a variety of procedures that enable you to produce reports containing data output. This chapter explains how to create and print reports using the PRINT procedure, one of the most commonly used report-writing procedures. The following sections also explain how to produce reports containing selected variables, observations, and data groups.

Producing Reports

The SAS System enables you to produce printed reports of computer output with the PRINT procedure. This procedure prints output of a SAS data set in an easy-to-read form. Producing reports is important because it enables you to present a detailed picture of a SAS data set. You also can use reports as references when you're analyzing data or to verify that the data were entered correctly.

The following statements produce the report illustrated in Output 7.1:

```
proc print data=htwt;
   title 'Height-Weight Study';
   footnote5 '1990 Data';
run;
```

Output 7.1
PROC PRINT
Report

```
                         Height-Weight Study                              1

          OBS    NAME        SEX    AGE    HEIGHT    WEIGHT

           1     Aubrey       M      41      74       170
           2     Ron          M      42      68       166
           3     Carl         M      32      70       155
           4     Antonio      M      39      72       167
           5     Deborah      F      30      66       124
           6     Jacqueline   F      33      66       115
           7     Helen        F      26      64       121
           8     David        M      30      71       158
           9     James        M      53      72       175
          10     Michael      M      32      69       143
          11     Ruth         F      47      69       139
          12     Joel         M      34      72       163

                                        (continued on next page)
```

(continued from previous page)

13	Donna	F	23	62	98
14	Roger	M	36	75	160
15	Yao	M	.	70	145
16	Elizabeth	F	31	67	135
17	Tim	M	29	71	176
18	Susan	F	28	65	131

1990 Data

Note: Remember that TITLE and FOOTNOTE statements can be combined with SAS procedures. See "Adding and Removing Titles and Footnotes" in Chapter 5, "Using SAS Procedures," for more information.

Now that you are more familiar with PROC PRINT, let's review a few statements that can be combined with PROC PRINT to create enhanced reports.

Selecting and Ordering Variables Using the VAR Statement

The SAS System enables you to process selected variables and specify the order in which you want them to appear with the VAR statement mentioned in Chapter 5. The VAR statement can be combined with PROC PRINT and other SAS procedures. For example, the following statements produce a report containing the NAME, AGE, SEX, and HEIGHT variables, as illustrated in Output 7.2:

```
proc print data=htwt;
   var name age sex height;
run;
```

Output 7.2
PROC PRINT with
VAR Statement

		The SAS System			1
OBS	NAME	AGE	SEX	HEIGHT	
1	Aubrey	41	M	74	
2	Ron	42	M	68	
3	Carl	32	M	70	
4	Antonio	39	M	72	
5	Deborah	30	F	66	
6	Jacqueline	33	F	66	
7	Helen	26	F	64	
8	David	30	M	71	
9	James	53	M	72	
10	Michael	32	M	69	
11	Ruth	47	F	69	
12	Joel	34	M	72	
13	Donna	23	F	62	
14	Roger	36	M	75	
15	Yao	.	M	70	
16	Elizabeth	31	F	67	
17	Tim	29	M	71	
18	Susan	28	F	65	

Notice that the SAS System prints only the variables specified and lists them in the order they appear in the VAR statement.

Selecting Observations Using the WHERE Statement

The WHERE statement also can be used with PROC PRINT and other procedures to select specified observations in a data set. As you learned in Chapter 5, the SAS System processes observations that meet certain conditions specified in the WHERE statement. For example, the following statements produce a report containing a list of women weighing more than 120 pounds:

```
proc print data=htwt;
   var sex weight;
   where sex='F' and weight>120;
run;
```

Printing Data in Groups Using BY Statements

The SAS System also enables you to manipulate groups of data to produce the results you need. For example, suppose you want to produce separate lists of the males and females in the height-weight study. To print a list of each group, sort the data using the SORT procedure and specify the sort order using a BY statement. Remember that observations must be sorted in the order of the variables specified in the BY statement before you can use a BY statement with any SAS procedure. The following statements illustrate this process:

```
proc sort data=htwt out=sorthtwt;
   by sex;
run;

proc print data=sorthtwt;
   by sex;
run;
```

The resulting data are grouped by the values of the variable SEX, as illustrated in Output 7.3.

```
                              The SAS System                              1
---------------------------------- SEX=F ----------------------------------

        OBS    NAME         AGE    HEIGHT    WEIGHT

         1     Deborah       30      66       124
         2     Jacqueline    33      66       115
         3     Helen         26      64       121
         4     Ruth          47      69       139
         5     Donna         23      62        98
         6     Elizabeth     31      67       135
         7     Susan         28      65       131


---------------------------------- SEX=M ----------------------------------

        OBS    NAME         AGE    HEIGHT    WEIGHT

         8     Aubrey        41      74       170
         9     Ron           42      68       166
        10     Carl          32      70       155
        11     Antonio       39      72       167
        12     David         30      71       158
        13     James         53      72       175
        14     Michael       32      69       143
        15     Joel          34      72       163
        16     Roger         36      75       160
        17     Yao            .      70       145
        18     Tim           29      71       176
```

The system also enables you to break your data into smaller groups by using two or more variables in the BY statement. For example, you can print a list of 30-year-old females, 30-year-old males, 31-year-old females, and so on by submitting the following statements:

```
proc sort data=htwt out=sorthtwt;
   by age sex;
run;

proc print data=sorthtwt;
   by age sex;
run;
```

Output 7.4 shows the first page of the results.

Output 7.4
Printing with Two BY Variables

```
                              The SAS System                            1
-------------------------------- AGE=. SEX=M --------------------------------

          OBS     NAME     HEIGHT     WEIGHT

           1      Yao        70         145


-------------------------------- AGE=23 SEX=F -------------------------------

          OBS     NAME     HEIGHT     WEIGHT

           2      Donna      62         98


-------------------------------- AGE=26 SEX=F -------------------------------

          OBS     NAME     HEIGHT     WEIGHT

           3      Helen      64         121


-------------------------------- AGE=28 SEX=F -------------------------------

          OBS     NAME     HEIGHT     WEIGHT

           4      Susan      65         131


-------------------------------- AGE=29 SEX=M -------------------------------

          OBS     NAME     HEIGHT     WEIGHT

           5      Tim        71         176


-------------------------------- AGE=30 SEX=F -------------------------------

          OBS     NAME     HEIGHT     WEIGHT

           6      Deborah    66         124


-------------------------------- AGE=30 SEX=M -------------------------------

          OBS     NAME     HEIGHT     WEIGHT

           7      David      71         158


-------------------------------- AGE=31 SEX=F -------------------------------

          OBS     NAME     HEIGHT     WEIGHT

           8      Elizabeth  67         135
```

Chapter 8 Plotting Data Using the PLOT Procedure

Introduction

This chapter describes the PLOT procedure. PROC PLOT enables you to describe the relationship between variables by plotting their values. The following sections explain how to create plots and control their appearance by changing plotting symbols and specifying tick mark locations.

Plotting Data

Plotting data enables you to graphically illustrate the relationship between variables. The PLOT procedure reads the values for sets of variables and marks the intersection of each pair of values as a point on the plot.

To produce a simple plot of one set of variables, enter a PROC PLOT statement followed by a PLOT statement. The PROC PLOT statement instructs the SAS System to produce a plot using specified data. PLOT statements specify the variables you want plotted. The PLOT statement begins with the keyword PLOT and is followed by the names of the variables you want plotted on the vertical and horizontal axes, joined by an asterisk.

For example, the following statements produce a plot of height by weight, with the variable HEIGHT on the vertical axis and the variable WEIGHT on the horizontal axis:

```
proc plot data=htwt;
   plot height*weight;
run;
```

Figure 8.1 shows the results.

Figure 8.1
PROC PLOT
Output

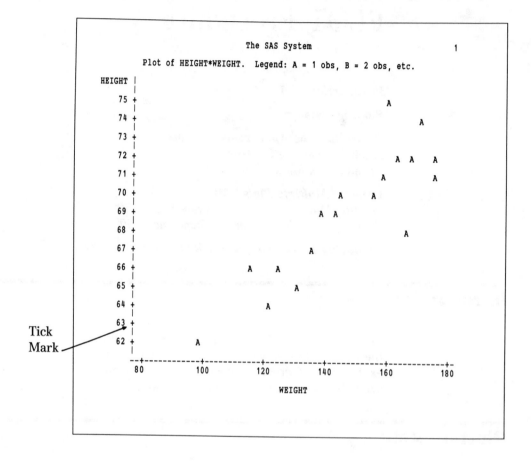

Tick
Mark

As Figure 8.1 illustrates, PROC PLOT automatically selects the plotting symbol A to represent one occurrence at each point. If two occurrences coincide at a point, the plotting symbol B is used; if three coincide, the symbol C is used; and so forth. PROC PLOT also selects ranges for both axes, places tick marks at reasonably spaced intervals, and prints a legend that names the variables and explains the plotting symbols.*

Controlling the Appearance of Plots

Although PROC PLOT determines many plot characteristics by default, such as the type of plotting symbol or the tick mark location, the SAS System enables you to control the appearance of your plot by overriding the default characteristics and specifying selections of your own. The following sections describe how to enhance plots to increase their effectiveness.

---

* All output in this chapter is created with the PAGESIZE= option set to 40 and the LINESIZE= option set to 76 in the OPTIONS statement. Remember that these options remain in effect until you reset them or end the SAS session.

Specifying Plotting Symbols

Although PROC PLOT uses letters of the alphabet as default plotting symbols, you can override the system default and specify plotting symbols of your choice. To specify a plotting symbol, enter a PLOT statement containing the vertical and horizontal variable names, as shown earlier, followed by an equal sign (=) and the character you select surrounded by single or double quotes. For example, the PLOT statement in the following example specifies an asterisk as the plotting symbol:

```
proc plot data=htwt;
   plot age*height='*';
run;
```

Output 8.1 shows the results.

Output 8.1
Specifying a
Plotting Symbol

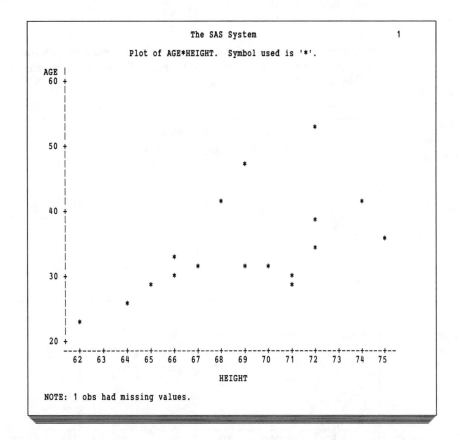

Note: When you specify a plotting symbol, PROC PLOT uses that symbol for all points on the plot regardless of how many points coincide. If points coincide, a message appears at the bottom of the plot telling you how many observations are hidden.

Also notice how the SAS System handles missing values. The note at the bottom of the page informs you that one observation contains a missing value.

Defining Tick Marks

The SAS System also enables you to control the location of tick marks on the horizontal axis by specifying your selection using the HAXIS= option in the PLOT statement. A corresponding VAXIS= option controls tick mark location on the vertical axis.

Specify your tick mark location in the PLOT statement by entering the variable names followed by a slash, the option name, an equal sign, and then the values you want to assign to the tick marks. To specify the tick marks on both axes, enter the HAXIS= and VAXIS= options separated by one or more blank spaces, as follows:

```
proc plot data=htwt;
    plot age*weight / haxis=80 100 120 140 160 180 200
                      vaxis=20 to 65 by 5;
    title 'Height-Weight Study';
    footnote '1990 Data';
run;
```

Notice that you can list the tick mark values separately, as the HAXIS= option illustrates, or you can abbreviate them, as illustrated with the VAXIS= option.

Output 8.2 shows the results.

Output 8.2
Specifying Tick
Mark Locations

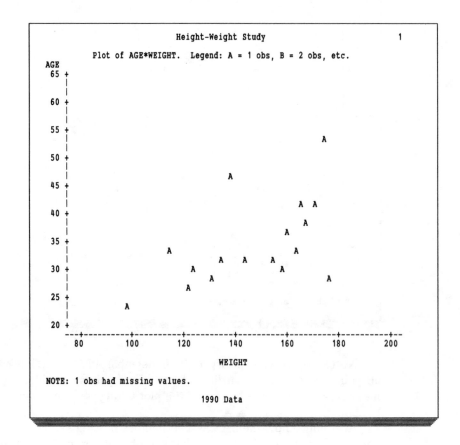

Note: As Output 8.2 illustrates, you can add titles and footnotes to your plots with the TITLE and FOOTNOTE statements. You also can plot specific observations or subsets of data with the WHERE and BY statements. Refer to Chapter 5, "Using SAS Procedures," for more information on these statements.

Creating Multiple Plots

The following sections explain how to create multiple plots in a variety of formats.

Creating Multiple Plots on Separate Pages

The SAS System enables you to produce multiple plots from the same data set by specifying additional sets of variables in the PLOT statement. For example, the following statements produce two plots on separate pages:

```
proc plot data=htwt;
   plot age*height='x' age*weight='o';
run;
```

Output 8.3 shows the results.

Output 8.3
Producing
Multiple Plots

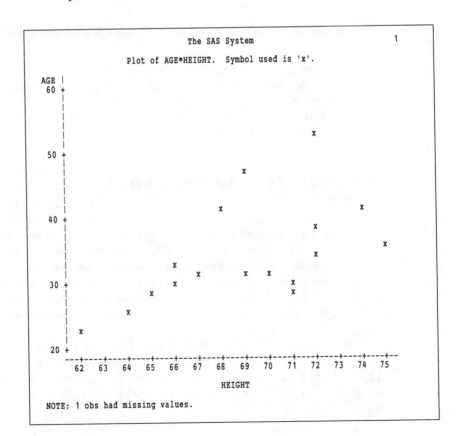

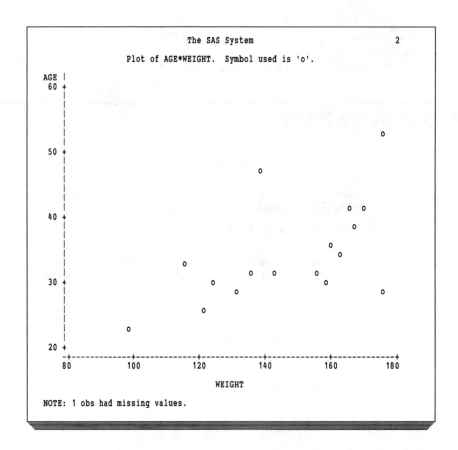

```
                              The SAS System                              2
                  Plot of AGE*WEIGHT.  Symbol used is 'o'.

   AGE |
    60 +
       |
       |
       |
       |
       |                                                              o
    50 +
       |
       |                                    o
       |
       |
    40 +                                               o   o
       |                                                o
       |                                          o
       |                                        o
    30 +               o              o   o        o
       |                   o                     o
       |                 o
       |             o
       |        o
    20 +
      -+------------+------------+------------+------------+------------+
       80          100          120          140          160          180
                                    WEIGHT

NOTE: 1 obs had missing values.
```

Notice that each plot uses a different plotting symbol.

To superimpose multiple plots on one set of axes, use the OVERLAY option in the PLOT statement. The OVERLAY option is described in detail in *SAS Language and Procedures: Usage, Version 6, First Edition* and *SAS Procedures Guide, Version 6, Third Edition*. Refer to these books for more information.

Creating Multiple Plots on the Same Page

PROC PLOT provides two options that produce two or more plots on the same page: the VPERCENT= and HPERCENT= options. You can use the VPERCENT= option to produce multiple plots on a page, with one plot beneath the other. The corresponding HPERCENT= option enables you to place one or more plots side by side on a page. The numeric value assigned to the option specifies the percentage of the vertical or horizontal dimension of the page to assign to each plot.

Specify the VPERCENT= and HPERCENT= options in the PROC PLOT statement by entering the option name, an equal sign, and the number representing the percentage of the page that you want the plot to occupy. For example, to produce two plots on a page, with one beneath the other, submit the following statements:

```
proc plot data=htwt vpercent=50;
run;
```

Figure 8.2 illustrates the format of the results.

Figure 8.2
PROC PLOT with
the VPERCENT=
Option

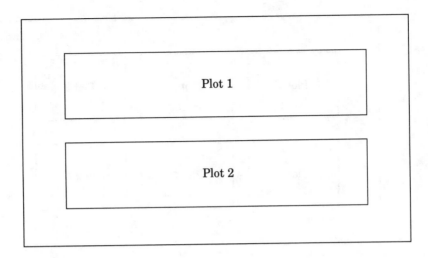

To produce three plots on a page, side by side, submit the following statements:

```
proc plot data=htwt hpercent=33;
run;
```

Figure 8.3 illustrates the format of the results.

Figure 8.3
PROC PLOT with
the HPERCENT=
Option

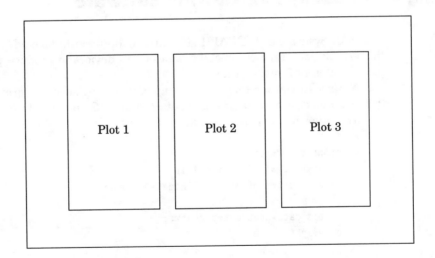

To produce six plots on a page, use the VPERCENT= and HPERCENT= options together, as follows:

```
proc plot data=htwt hpercent=33 vpercent=50;
run;
```

Figure 8.4 illustrates the format of the results.

Figure 8.4
PROC PLOT with
the HPERCENT=
and VPERCENT=
Options

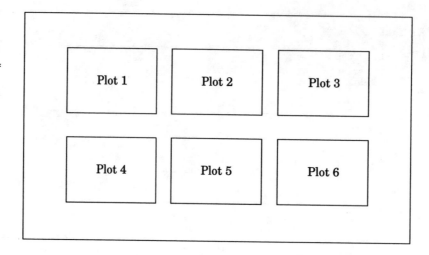

You also can use the abbreviations VPCT= and HPCT= for these options, as follows:

```
proc plot hpct=33 vpct=50;
run;
```

Producing Plots Using SAS/GRAPH Software

The SAS System's SAS/GRAPH software enables you to take advantage of the high-resolution graphics capabilities of output devices to produce plots that include color, different fonts, and text.

Note: To run the following example, you must specify a graphics device. If you are not sure how to do this, contact your SAS Software Consultant.

The following statements produce Output 8.4:

```
goptions gunit=pct
         hsize=10 in vsize=7 in
         ftext=zapf htitle=6 htext=3 border;

proc sort data=htwt out=sorthtwt;
   by weight;
run;

proc gplot data=sorthtwt;
   plot height*weight / frame;
   symbol1 value=dot h=2 i=join;
   title 'Plot of HEIGHT versus WEIGHT';
   title2 h=4 'Using the GPLOT Procedure';
run;
quit;
```

Output 8.4   *Producing SAS/GRAPH Plots*

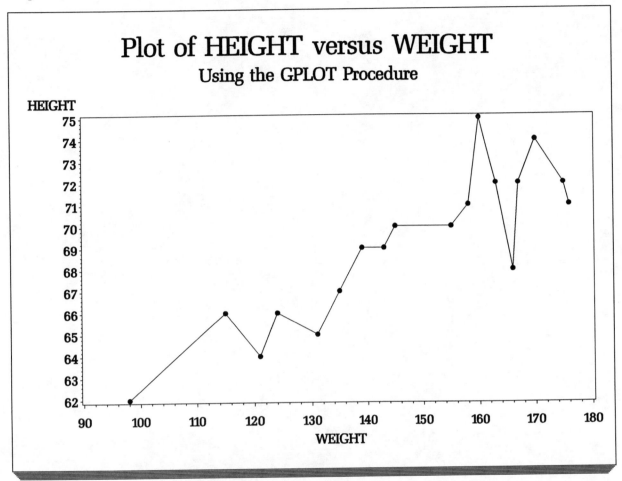

Refer to *SAS/GRAPH Software: Introduction, Version 6, First Edition* or
SAS/GRAPH Software: Reference, Version 6, First Edition, Volume 1 and *Volume 2*
for complete documentation on using SAS/GRAPH software.

64

Chapter 9 Charting Data Using the CHART Procedure

Introduction

This chapter describes the CHART procedure and explains how to use the VBAR, HBAR, BLOCK, and PIE statements to produce vertical and horizontal bar charts, block charts, and pie charts. In addition, you will learn how to control the appearance of your charts using statement options.

Charting Data

The CHART procedure produces various charts, including vertical and horizontal bar charts, block charts, pie charts, and star charts. The following sections describe the most commonly used charts: bar charts, block charts, and pie charts.

Creating Vertical Bar Charts

The SAS System enables you to create vertical bar charts using PROC CHART and VBAR statements. The PROC CHART statement instructs the system to produce a chart using specified data. The VBAR statement specifies the type of chart and the variable name. In the following example, the VBAR statement begins with the keyword VBAR and is followed by the name of the variable whose values you are charting. This variable is often referred to as the *chart variable*.

```
proc chart data=htwt;
   vbar height;
run;
```

Output 9.1 shows the results.*

Output 9.1
Vertical Bar Chart:
Numeric Variable

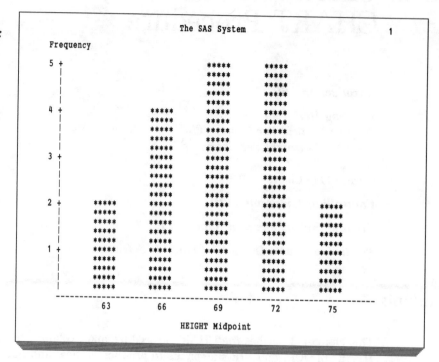

```
                              The SAS System                              1

      Frequency

        5 +                         *****     *****
          |                         *****     *****
          |                         *****     *****
          |                         *****     *****
          |                         *****     *****
        4 +               *****     *****     *****
          |               *****     *****     *****
          |               *****     *****     *****
          |               *****     *****     *****
          |               *****     *****     *****
        3 +               *****     *****     *****
          |               *****     *****     *****
          |               *****     *****     *****
          |               *****     *****     *****
          |               *****     *****     *****
        2 +     *****      *****     *****     *****     *****
          |     *****      *****     *****     *****     *****
          |     *****      *****     *****     *****     *****
          |     *****      *****     *****     *****     *****
          |     *****      *****     *****     *****     *****
        1 +     *****      *****     *****     *****     *****
          |     *****      *****     *****     *****     *****
          |     *****      *****     *****     *****     *****
          |     *****      *****     *****     *****     *****
          |     *****      *****     *****     *****     *****
          -------------------------------------------------------
                  63        66        69        72        75

                             HEIGHT Midpoint
```

Output 9.1 is a frequency chart of the variable HEIGHT. *Frequency charts*
graphically illustrate *frequency counts*, the number of times a value or range of
values occurs in a given data set.** PROC CHART creates frequency charts by
default. The size of each bar, block, or section in a frequency chart represents the
number of values falling within a specific range.

As Output 9.1 illustrates, the system automatically labels both axes. It also
places tick marks on the vertical axis and selects midpoints for the horizontal axis.
Midpoints represent the center of a range of values.

For example, the midpoints for the variable HEIGHT range from 63 to 75
inches at intervals of 3 inches. The midpoint of 63 inches represents the center of
a range of 61.5 inches up to but not including 64.5 inches. A midpoint of 66
inches represents a range from 64.5 inches up to but not including 67.5 inches,
and so forth.

---

* All output in this chapter is created with the PAGESIZE= option set to 40 and the LINESIZE= option
   set to 76 in the OPTIONS statement, unless otherwise stated. Once you set the PAGESIZE= and
   LINESIZE= options, they remain in effect until you reset them or end the SAS session.

** All the charts in this chapter are frequency charts. For information on other types of charts, refer to
   SAS Language and Procedures: Usage, Version 6, First Edition.

If you are charting a character variable, such as the variable SEX in the following example, PROC CHART does not use ranges. Instead, it creates one bar, block, or section for each value of the chart variable.

```
proc chart data=htwt;
   vbar sex;
run;
```

Output 9.2 shows the results.

Output 9.2
Vertical Bar Chart:
Character Variable

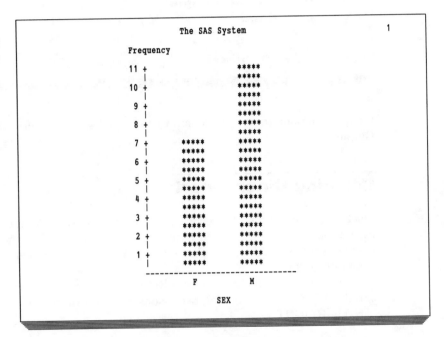

Creating Horizontal Bar Charts

Horizontal bar charts are similar to vertical bar charts, except the horizontal axis represents the frequency and the vertical axis represents the variable. Horizontal bar charts also contain additional statistics, such as the frequency, cumulative frequency, percent, and cumulative percent.

Frequency is the number of observations in a given range. *Cumulative frequency* is the number of observations in all ranges up to and including a given range. The cumulative frequency for the last range is equal to the number of observations in the data set.

Percent is the percentage of observations in a given range. *Cumulative percent* is the percentage of observations in all ranges up to and including a given range. The cumulative percent for the last range is always 100.

To create a horizontal bar chart, use the HBAR statement, which specifies the variable you want charted. For example, the following statements create a horizontal bar chart of the variable HEIGHT:

```
proc chart data=htwt;
   hbar height;
run;
```

Output 9.3 shows the results.

Output 9.3
Horizontal Bar
Chart

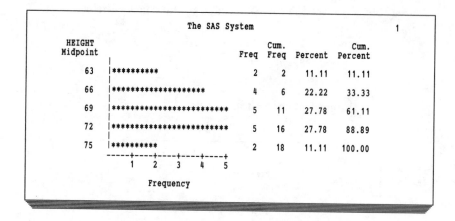

Compare the horizontal bar chart in Output 9.3 with the vertical bar chart in
Output 9.1.

Creating Block Charts

Block charts are similar to vertical bar charts because they emphasize individual
ranges, but block charts produce more polished output. However, block charts
may be less precise than bar charts because blocks are limited to a maximum
height of ten lines.

To create a block chart, enter a BLOCK statement specifying the chart
variable. For example, the following statements create a block chart of the
variable HEIGHT:

```
proc chart data=htwt;
   block height;
run;
```

Output 9.4 shows the results.*

Output 9.4
Block Chart

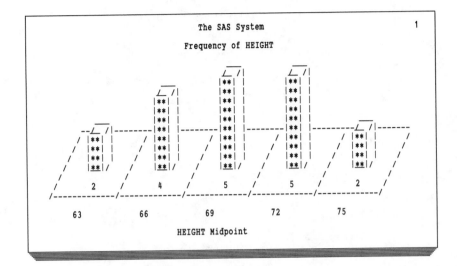

As Output 9.4 illustrates, the resulting block chart contains five blocks representing various ranges of values. The first block represents values ranging from 61.5 inches up to but not including 64.5 inches. The midpoint for this range is 63 inches. As you can see, two observations fall within this range. Four observations fall within the second range, five within the third range, and so forth.

Notice that the SAS System automatically places a second header in Output 9.4. This second default header, which appears in block and pie charts, includes the name of the chart variable. You can override this default header with the NOHEADER option. Refer to *SAS Procedures Guide, Version 6, Third Edition* for more information on the NOHEADER option.

Note: If the line size or page size is insufficient for the procedure to create a block chart, PROC CHART creates a horizontal bar chart instead. The procedure sends a message to your SAS log explaining what happened.

Creating Pie Charts

Pie charts also illustrate the distribution of a variable's values but emphasize the relationship of each range to the whole. Pie charts are created using the PIE statement, which specifies the chart variable. For example, the following statements create a pie chart of the variable HEIGHT:

```
proc chart data=htwt;
   pie height;
run;
```

Output 9.5 shows the results.

---

* This block chart was generated with the LINESIZE= option set to 80 in the OPTIONS statement. Remember that this option remains in effect until you reset it or end the current SAS session.

Output 9.5
Pie Chart

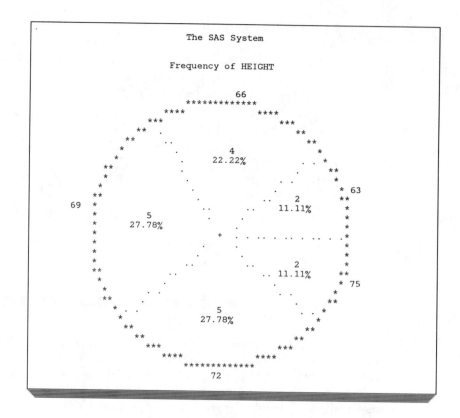

The number of sections in a pie chart corresponds to the number of bars in a vertical bar chart, with one exception. When creating pie charts, PROC CHART combines all very small sections into one section called other. The size of a section corresponds to the number of observations falling in its range.

As Output 9.5 illustrates, the majority of values for the variable HEIGHT fall within two ranges, one with a midpoint of 69 inches and the other with a midpoint of 72 inches. Both of these ranges contain five observations, or 27.78 percent of the total number of observations. Note that the output also contains a default title and header.

Controlling Midpoints

As mentioned earlier in this chapter, the CHART procedure selects many chart characteristics by default. For example, PROC CHART automatically labels axes and selects ranges for numeric variables. You can override these default characteristics if you want to change the appearance of the chart.

For example, you can specify midpoints using the MIDPOINTS= option. Midpoints, as you may recall, represent the center value in a range of values. The following example illustrates the MIDPOINTS= option:

```
proc chart data=htwt;
   vbar height / midpoints=60 65 70 75;
run;
```

Output 9.6 shows the results.

Output 9.6
Specifying
Midpoints

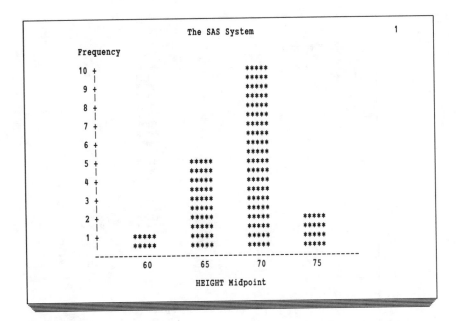

Refer to *SAS Procedures Guide, Version 6, Third Edition* for more information on midpoints.

Creating Subgroups in a Range

The SAS System enables you to chart subsets of your data with various options. The SUBGROUP= option illustrates how subgroups contribute to each bar or block in a chart.* For example, the SUBGROUP= option is useful if you want to illustrate how many observations in a group represent males and how many represent females.

The SUBGROUP= option defines a variable called the *subgroup variable*. PROC CHART creates a set of bars or blocks for each value of the subgroup variable. PROC CHART fills each bar or block with characters that show the contribution of each value of the subgroup variable to the total. For example, the following program defines the variable SEX as the subgroup variable:

```
proc chart data=htwt;
   vbar height / midpoints=60 65 70 75
                 subgroup=sex;
run;
```

The results in Output 9.7 illustrate the number of males and females composing each bar.

---

* The SUBGROUP= option cannot be used in the PIE statement.

Output 9.7
Subgrouping Data

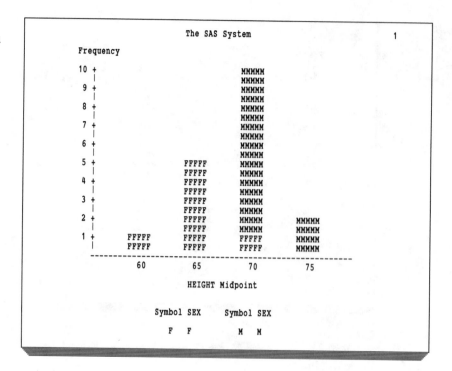

```
                              The SAS System                          1

     Frequency

      10 +                                 MMMMM
         |                                 MMMMM
       9 +                                 MMMMM
         |                                 MMMMM
       8 +                                 MMMMM
         |                                 MMMMM
       7 +                                 MMMMM
         |                                 MMMMM
       6 +                                 MMMMM
         |                                 MMMMM
       5 +                     FFFFF       MMMMM
         |                     FFFFF       MMMMM
       4 +                     FFFFF       MMMMM
         |                     FFFFF       MMMMM
       3 +                     FFFFF       MMMMM
         |                     FFFFF       MMMMM
       2 +                     FFFFF       MMMMM       MMMMM
         |                     FFFFF       MMMMM       MMMMM
       1 +         FFFFF       FFFFF       FFFFF       MMMMM
         |         FFFFF       FFFFF       FFFFF       MMMMM
           ------------------------------------------------------
                    60          65          70          75

                              HEIGHT Midpoint


              Symbol SEX        Symbol SEX

                 F    F            M    M
```

Notice that each bar contains a proportional amount of characters representing the values of the variable SEX. For example, the first two bars in Output 9.7 consist of characters representing only females. The third bar consists of characters representing both males and females, and the fourth bar represents only males.

In some cases, you may find it more effective to create a separate bar for each value of a variable. The GROUP= option defines a variable called the *group variable* and creates a set of bars or blocks for each value of this variable.* The following statements produce the chart in Output 9.8:

```
proc chart data=htwt;
    vbar height / midpoints=60 65 70 75
                  group=sex;
run;
```

---

*   The PIE statement doesn't support this option.

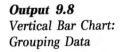

Output 9.8
Vertical Bar Chart:
Grouping Data

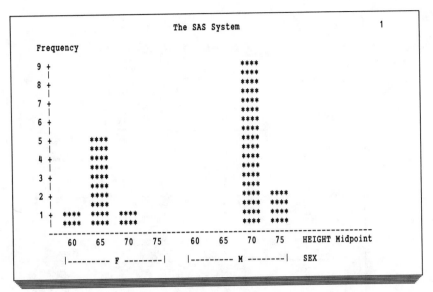

Output 9.8 includes a set of bars for each value of the group variable SEX. Notice that each set of bars uses the midpoints you specify with the MIDPOINTS= option.

Complicated relationships like those charted with the GROUP= option are easier to understand if you present them as block charts because the resulting chart is three-dimensional. To create a three-dimensional chart, change the VBAR statement to a BLOCK statement as in the following, and adjust the line size if necessary:

```
proc chart data=htwt;
   block height / midpoints=60 65 70 75
                  group=sex;
run;
```

Output 9.9 shows the results.

Output 9.9
Block Chart:
Grouping Data

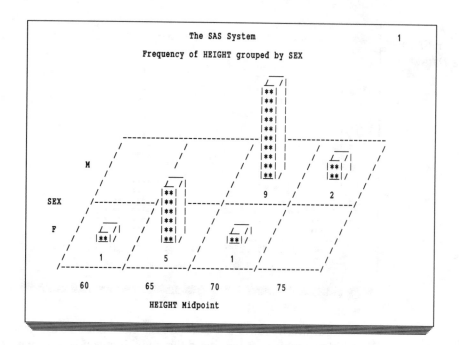

Producing Charts Using SAS/GRAPH Software

The SAS System's SAS/GRAPH software enables you to take advantage of the high-resolution graphics capabilities of output devices to produce charts that include color, different fonts, and text.

Note: To run the following example, you must specify a graphics device. If you are not sure how to do this, contact your SAS Software Consultant. Refer to *SAS/GRAPH Software: Introduction, Version 6, First Edition* or *SAS/GRAPH Software: Reference, Version 6, First Edition, Volume 1* and *Volume 2* for complete documentation on using SAS/GRAPH software.

The following statements produce Output 9.10:

```
goptions gunit=pct
         hsize=10 in vsize=7 in
         ftext=zapf htitle=6 htext=3 border;

proc gchart data=htwt;
   block height / midpoints=60 65 70 75
                  group=sex
                  patternid=group;
   pattern1 color=blue value=x1;
   pattern2 color=green value=x5;
   title 'Block Chart of HEIGHT Grouped by SEX';
run;
quit;
```

Block Chart of HEIGHT Grouped by SEX

FREQUENCY BLOCK CHART

The example produces output featuring various colors and patterns, although the colors are not visible in this book.

Chapter **10** Generating Frequency and Crosstabulation Tables Using the FREQ Procedure

Introduction

This chapter describes the FREQ procedure, which summarizes data in tabular format. PROC FREQ produces frequency tables and crosstabulation tables. These tables illustrate the frequency at which individual values or combinations of values occur within a SAS data set.

In addition, you will learn how to format tables using the following TABLES statement options: NOCUM, NOCOL, NOPERCENT, NOROW, and MISSING.

Producing Frequency Tables

The SAS System enables you to create frequency tables with the FREQ procedure. *Frequency tables* summarize data by displaying the frequency count, or how often the value of a variable occurs in a SAS data set. These tables also display the percent, cumulative frequency, and cumulative percent of each data value. Frequency tables that process one variable often are referred to as one-way tables.

For example, suppose you want to know how many 35-year-olds are in the height-weight study, how many 36-year-olds, and so on. To determine the number of observations with an AGE value of 35, 36, and so on, enter a PROC FREQ statement followed by a TABLES statement. The TABLES statement specifies the variable you want processed. Simply enter the keyword TABLES followed by the variable name, as follows:

```
proc freq data=htwt;
   tables age;
run;
```

Output 10.1 shows the results.

Output 10.1
Default Frequency
Table

```
                              The SAS System                              1

                                         Cumulative  Cumulative
          AGE   Frequency   Percent      Frequency    Percent
          ---------------------------------------------------------
           23       1         5.9            1          5.9
           26       1         5.9            2         11.8
           28       1         5.9            3         17.6
           29       1         5.9            4         23.5
           30       2        11.8            6         35.3
           31       1         5.9            7         41.2
           32       2        11.8            9         52.9
           33       1         5.9           10         58.8
           34       1         5.9           11         64.7
           36       1         5.9           12         70.6
           39       1         5.9           13         76.5
           41       1         5.9           14         82.4
           42       1         5.9           15         88.2
           47       1         5.9           16         94.1
           53       1         5.9           17        100.0

                        Frequency Missing = 1
```

Note: By default, missing values do not appear in frequency and crosstabulation tables like other values, but the system does include a count of missing values beneath the table. PROC FREQ includes missing values when it calculates the percentages for rows, columns, and totals only if the MISSING option is specified. This option instructs the SAS System to treat missing values like nonmissing values and include them in calculations of percentages and other statistics. For more information on this option, refer to *SAS Procedures Guide, Version 6, Third Edition*.

The TABLES statement also enables you to specify options that suppress specific categories of output. For example, suppose you want to generate a frequency table without the cumulative statistics. Simply specify the NOCUM option by entering a slash (/) and the option name NOCUM after the variable name in the TABLES statement, as follows:

```
proc freq data=htwt;
   tables age / nocum;
run;
```

Output 10.2 shows the results.

Output 10.2
Frequency Table
without Cumulative
Statistics

```
                        The SAS System                        1

              AGE   Frequency   Percent
              -------------------------------
               23        1        5.9
               26        1        5.9
               28        1        5.9
               29        1        5.9
               30        2       11.8
               31        1        5.9
               32        2       11.8
               33        1        5.9
               34        1        5.9
               36        1        5.9
               39        1        5.9
               41        1        5.9
               42        1        5.9
               47        1        5.9
               53        1        5.9

                   Frequency Missing = 1
```

As you can see in Output 10.2, the NOCUM option suppresses the display of the cumulative frequencies and cumulative percents that appear in Output 10.1.

Note: Use the NOPERCENT option in the TABLES statement to suppress the display of the percent column. Refer to *SAS Procedures Guide* for more information on this option.

Producing Crosstabulation Tables

The SAS System enables you to describe data further with a crosstabulation table. A *crosstabulation table* is a frequency table that displays the frequency distribution for two or more variables. Hence, these tables are often referred to as two-way, three-way, or *n*-way tables.

To produce a crosstabulation table, enter a TABLES statement containing the keyword TABLES, followed by the name of the variables you want to process, separated by an asterisk (*). The values of the first variable listed form the rows of the table; the values of the second variable form the columns.

For example, suppose you want to know how many 30-year-old females are in the height-weight study, how many 30-year-old males, how many 31-year-old females, and so on. Simply specify the two variables AGE and SEX in the TABLES statement, as follows:

```
proc freq data=htwt;
   tables sex*age;
run;
```

Output 10.3 shows the results.

Output 10.3
Default
Crosstabulation
Table

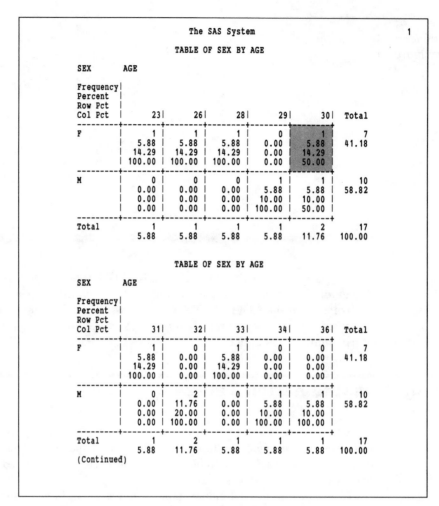

```
                              The SAS System                          1

                           TABLE OF SEX BY AGE

SEX        AGE

Frequency|
Percent  |
Row Pct  |
Col Pct  |     23|     26|     28|     29|     30| Total
---------+-------+-------+-------+-------+-------+
F        |      1|      1|      1|      0|      1|      7
         |   5.88|   5.88|   5.88|   0.00|   5.88|  41.18
         |  14.29|  14.29|  14.29|   0.00|  14.29|
         | 100.00| 100.00| 100.00|   0.00|  50.00|
---------+-------+-------+-------+-------+-------+
M        |      0|      0|      0|      1|      1|     10
         |   0.00|   0.00|   0.00|   5.88|   5.88|  58.82
         |   0.00|   0.00|   0.00|  10.00|  10.00|
         |   0.00|   0.00|   0.00| 100.00|  50.00|
---------+-------+-------+-------+-------+-------+
Total           1       1       1       1       2     17
             5.88    5.88    5.88    5.88   11.76  100.00


                           TABLE OF SEX BY AGE

SEX        AGE

Frequency|
Percent  |
Row Pct  |
Col Pct  |     31|     32|     33|     34|     36| Total
---------+-------+-------+-------+-------+-------+
F        |      1|      0|      1|      0|      0|      7
         |   5.88|   0.00|   5.88|   0.00|   0.00|  41.18
         |  14.29|   0.00|  14.29|   0.00|   0.00|
         | 100.00|   0.00| 100.00|   0.00|   0.00|
---------+-------+-------+-------+-------+-------+
M        |      0|      2|      0|      1|      1|     10
         |   0.00|  11.76|   0.00|   5.88|   5.88|  58.82
         |   0.00|  20.00|   0.00|  10.00|  10.00|
         |   0.00| 100.00|   0.00| 100.00| 100.00|
---------+-------+-------+-------+-------+-------+
Total           1       2       1       1       1     17
             5.88   11.76    5.88    5.88    5.88  100.00
(Continued)
```

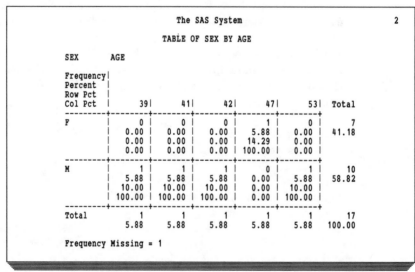

```
                              The SAS System                          2

                           TABLE OF SEX BY AGE

SEX        AGE

Frequency|
Percent  |
Row Pct  |
Col Pct  |     39|     41|     42|     47|     53| Total
---------+-------+-------+-------+-------+-------+
F        |      0|      0|      0|      1|      0|      7
         |   0.00|   0.00|   0.00|   5.88|   0.00|  41.18
         |   0.00|   0.00|   0.00|  14.29|   0.00|
         |   0.00|   0.00|   0.00| 100.00|   0.00|
---------+-------+-------+-------+-------+-------+
M        |      1|      1|      1|      0|      1|     10
         |   5.88|   5.88|   5.88|   0.00|   5.88|  58.82
         |  10.00|  10.00|  10.00|   0.00|  10.00|
         | 100.00| 100.00| 100.00|   0.00| 100.00|
---------+-------+-------+-------+-------+-------+
Total           1       1       1       1       1     17
             5.88    5.88    5.88    5.88    5.88  100.00

Frequency Missing = 1
```

Output 10.3 is a default crosstabulation table containing frequency, percent, row percent, and column percent statistics. The table consists of blocks of data, also known as *cells*, where the rows and columns intersect. Each cell contains four statistics for variables with specific values.

For example, the shaded cell contains data for all females with an AGE value of 30. The first number in the cell is the frequency, or the number of times the value occurs in the data set. The second number is the percent, or the percentage of observations represented by the number. The third number is the *row percent*, or the percentage of observations represented in that row of output. The last statistic provided, *column percent*, is the percentage of observations represented in the column.

As Output 10.3 illustrates, the HTWT data set contains only one observation with a value of **F** for the variable SEX and a value of 30 for the variable AGE. This observation represents 5.88 percent of the total number of observations in the data set, 14.29 percent of the total number of observations in that row (that is, for all observations where the value of variable SEX is **F**, regardless of age), and 50.00 percent of the total number of observations in that column (that is, for all observations where the value of variable AGE is 30, regardless of sex).

The SAS System also enables you to create *n*-way tables by connecting all variable names in the TABLES statement with asterisks. For example, the following statements generate a three-way table based on SEX, AGE, and HEIGHT. To simplify the PROC FREQ output, the DATA step creates a new data set, HTWT1, and condenses all the values of the variables HEIGHT and AGE to two values for each variable: **average** and **tall** for the new variable HEIGHT2 and **group1** and **group2** for the new variable AGE2. PROC FREQ then analyzes the data in this data set and produces the results in Output 10.4.

```
data htwt1;
   set htwt;
   if height<74 then height2='average';
   else height2='tall';
   if age<40 then age2='group1';
   else age2='group2';
run;

proc freq data=htwt1;
   tables sex*age2*height2;
run;
```

Output 10.4
Three-Way
Crosstabulation
Table

```
                          The SAS System                          1
                     TABLE 1 OF AGE2 BY HEIGHT2
                       CONTROLLING FOR SEX=F

            AGE2       HEIGHT2

            Frequency|
            Percent  |
            Row Pct  |
            Col Pct  |average |tall    |  Total
            ---------+--------+--------+
            group1   |      6 |      0 |      6
                     |  85.71 |   0.00 |  85.71
                     | 100.00 |   0.00 |
                     |  85.71 |      . |
            ---------+--------+--------+
            group2   |      1 |      0 |      1
                     |  14.29 |   0.00 |  14.29
                     | 100.00 |   0.00 |
                     |  14.29 |      . |
            ---------+--------+--------+
            Total           7        0        7
                       100.00     0.00   100.00

                     TABLE 2 OF AGE2 BY HEIGHT2
                       CONTROLLING FOR SEX=M

            AGE2       HEIGHT2

            Frequency|
            Percent  |
            Row Pct  |
            Col Pct  |average |tall    |  Total
            ---------+--------+--------+
            group1   |      7 |      1 |      8
                     |  63.64 |   9.09 |  72.73
                     |  87.50 |  12.50 |
                     |  77.78 |  50.00 |
            ---------+--------+--------+
            group2   |      2 |      1 |      3
                     |  18.18 |   9.09 |  27.27
                     |  66.67 |  33.33 |
                     |  22.22 |  50.00 |
            ---------+--------+--------+
            Total           9        2       11
                        81.82    18.18   100.00
```

Output 10.4 illustrates how many females and then how many males fall into each of the cells created by combining all the values of AGE2 and HEIGHT2.

If you do not need the default statistics illustrated in Output 10.4, use the NOCOL and NOROW options to suppress the display of this information. To suppress column percentages in your output, specify the NOCOL option in the TABLES statement by adding a slash (/) and the option name after the variable name. To suppress row percentages, specify the NOROW option. Both options are illustrated in Output 10.5.

```
proc freq data=htwt;
   tables age*sex / nocol norow;
run;
```

Output 10.5
*Crosstabulation
Table with NOCOL
and NOROW
Options*

```
                          The SAS System                          1

                       TABLE OF AGE BY SEX

           AGE      SEX

           Frequency|
           Percent  |F       |M       |  Total
           ---------+--------+--------+
               23   |   1 |   |   0 |      1
                    | 5.88 |  | 0.00 |   5.88
           ---------+--------+--------+
               26   |   1 |   |   0 |      1
                    | 5.88 |  | 0.00 |   5.88
           ---------+--------+--------+
               28   |   1 |   |   0 |      1
                    | 5.88 |  | 0.00 |   5.88
           ---------+--------+--------+
               29   |   0 |   |   1 |      1
                    | 0.00 |  | 5.88 |   5.88
           ---------+--------+--------+
               30   |   1 |   |   1 |      2
                    | 5.88 |  | 5.88 |  11.76
           ---------+--------+--------+
               31   |   1 |   |   0 |      1
                    | 5.88 |  | 0.00 |   5.88
           ---------+--------+--------+
               32   |   0 |   |   2 |      2
                    | 0.00 |  |11.76 |  11.76
           ---------+--------+--------+
               33   |   1 |   |   0 |      1
                    | 5.88 |  | 0.00 |   5.88
           ---------+--------+--------+
               34   |   0 |   |   1 |      1
                    | 0.00 |  | 5.88 |   5.88
           ---------+--------+--------+
               36   |   0 |   |   1 |      1
                    | 0.00 |  | 5.88 |   5.88
           ---------+--------+--------+
               39   |   0 |   |   1 |      1
                    | 0.00 |  | 5.88 |   5.88
           ---------+--------+--------+
               41   |   0 |   |   1 |      1
                    | 0.00 |  | 5.88 |   5.88
           ---------+--------+--------+
               42   |   0 |   |   1 |      1
                    | 0.00 |  | 5.88 |   5.88
           ---------+--------+--------+
               47   |   1 |   |   0 |      1
                    | 5.88 |  | 0.00 |   5.88
           ---------+--------+--------+
               53   |   0 |   |   1 |      1
                    | 0.00 |  | 5.88 |   5.88
           ---------+--------+--------+
           Total        7       10       17
                      41.18    58.82   100.00

           Frequency Missing = 1
```

Chapter 11 Generating Summary Statistics Using the MEANS Procedure

Introduction

This chapter describes the MEANS procedure. PROC MEANS summarizes data by calculating descriptive statistics for numeric variables. By default, PROC MEANS calculates the following statistics for all numeric variables in the input data set: the number of observations, the mean, the standard deviation, and the minimum and maximum values for each numeric variable.

Summarizing Data Using the MEANS Procedure

The MEANS procedure generates descriptive statistics for all numeric variables in a SAS data set. For example, the following program calculates the five default statistics for all numeric variables in the HTWT data set:

```
proc means data=htwt;
run;
```

Output 11.1 shows the results.

Output 11.1
Default PROC
MEANS Output

```
                              The SAS System                                1

        Variable   N        Mean        Std Dev       Minimum       Maximum
        -----------------------------------------------------------------------
        AGE        17    34.4705882     7.7630346    23.0000000    53.0000000
        HEIGHT     18    69.0555556     3.5225696    62.0000000    75.0000000
        WEIGHT     18   146.7222222    22.5409576    98.0000000   176.0000000
        -----------------------------------------------------------------------
```

Note: PROC MEANS does not include missing values when calculating statistics. For example, in Output 11.1 PROC MEANS uses only 17 of the 18 observations in the HTWT data set to calculate the mean for the variable AGE because the data set contains one observation with a missing value for AGE. If the variable AGE had no missing values, the procedure would use all 18 observations to calculate statistics for AGE.

To produce these statistics for some but not all numeric variables in a SAS data set, list the variables you want in a VAR statement. For example, the following program calculates the default statistics for the variables HEIGHT and WEIGHT only:

```
proc means data=htwt;
   var height weight;
run;
```

Generating Statistics Using Selected Keywords

The SAS System enables you to generate selected descriptive statistics using one or more keywords that correspond to PROC MEANS statistics. Table 11.1 lists some commonly used PROC MEANS keywords with the corresponding statistics they produce.

If you use the PROC MEANS keywords, you suppress the system default and must request each statistic you want.

Table 11.1
Keywords and the Statistics They Produce

Keyword	Statistic
N	the number of observations with nonmissing values of the variable
NMISS	the number of observations with missing values of the variable
MEAN	the mean or average
STD	the standard deviation
MIN	the minimum
MAX	the maximum
RANGE	the difference between the smallest and largest values
SUM	the sum of all the nonmissing values of the variable

Suppose you want to print only the number of observations and the mean of your numeric variables. Simply enter the keywords N and MEAN in the PROC MEANS statement, as follows:

```
proc means data=htwt n mean;
run;
```

Output 11.2 shows the results.

Output 11.2
PROC MEANS
Output

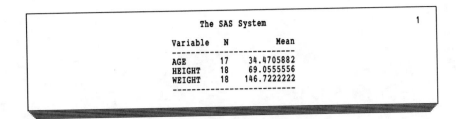

```
                        The SAS System                        1

              Variable   N        Mean
              ----------------------------------
              AGE       17    34.4705882
              HEIGHT    18    69.0555556
              WEIGHT    18   146.7222222
              ----------------------------------
```

Producing Group Statistics

You can produce descriptive statistics for groups or subsets of data by selecting specific variables using PROC MEANS and a BY statement. For example, to generate the mean for the variable WEIGHT first for females and then for males, enter the following statements:

```
proc sort data=htwt out=sorthtwt;
   by sex;
run;

proc means data=sorthtwt;
   by sex;
   var weight;
run;
```

Output 11.3 shows the results.

Output 11.3
PROC MEANS
Output Grouped
by SEX

```
                        The SAS System                        1

           Analysis Variable : WEIGHT

--------------------------------- SEX=F ---------------------------------

    N        Mean        Std Dev      Minimum      Maximum
    -----------------------------------------------------------
    7    123.2857143   13.8890159   98.0000000   139.0000000
    -----------------------------------------------------------

--------------------------------- SEX=M ---------------------------------

    N        Mean        Std Dev      Minimum      Maximum
    -----------------------------------------------------------
    11   161.6363636   10.9020432  143.0000000   176.0000000
    -----------------------------------------------------------
```

Chapter **12** Identifying Errors

Introduction

This chapter explains how the SAS System detects errors and shows you how to prevent common programming errors, understand SAS log messages, and verify that your data are correct.

How the SAS System Detects Errors

A SAS program is a series of one or more DATA or PROC steps. The SAS System processes programs by reading each statement separately, one word at a time in the order they appear. The system executes statements in the current step first and then continues with the next step. Any errors in the program are detected by the SAS System when it reaches a step boundary, such as a RUN statement or the next DATA or PROC statement.

When the system detects an error, it identifies the error's location and prints an explanation of the error in the SAS log. The SAS System detects three types of errors: syntax errors, execution-time errors, and data errors. These errors are described as follows:

☐ *Syntax errors* are errors made in SAS statements. Common syntax errors include misspelled keywords, missing semicolons, unmatched quotes, and other punctuation errors.

☐ *Execution-time errors* are errors detected when the system executes a step. For example, an execution-time error occurs if the system encounters an incorrect reference in an INFILE statement. Most serious execution-time errors cause the SAS System to stop executing the current step and write an error message to the SAS log. Execution of the program continues with the next step. Less serious errors generate messages from the SAS System but allow the step to run to completion.

☐ *Data errors* are errors that occur when raw data do not match values expected by the INPUT statement. A data error occurs, for example, if you specify numeric variables in the INPUT statement for character data. Data errors do not cause a step to stop but may cause the SAS System to write a message to the SAS log.

Preventing Common Programming Errors

To prevent errors, review each program before submitting it to the SAS System to process. The following list describes the most common errors in more detail and explains how to avoid them:

□   First, all SAS statements must appear in the correct order so the SAS System can execute them properly. Therefore, it's important to check the order of the statements in your program.

A DATA step must begin with a DATA statement. Other statements, such as IF-THEN statements, are located after the DATA statement. If you use the CARDS statement to indicate that input data lines follow, it must be the last statement before the data lines begin. In addition, there can be no statements between the CARDS statement and the end of the data lines. End each DATA step with a RUN statement to signal to the SAS System to execute the step.

The same principle applies to SAS PROC steps. A PROC step must begin with a PROC statement. Other statements, such as BY or VAR statements, are located after the PROC statement. End each PROC step with a RUN statement to signal to the SAS System to execute the step.

□   Once you have confirmed that all statements are in the proper order, check the syntax of the statements. For example, most SAS statements begin with a SAS keyword. Review each statement to determine whether you have omitted or misspelled any keywords or variable names.*

□   Also check that each statement ends with a semicolon. For example, if you omit the semicolon at the end of a PROC PRINT statement, the step does not execute, and the SAS System writes an error message to the SAS log.

□   In addition, check all single and double quotes to ensure that you have not omitted a starting or ending quote. If you omit a quote around a character string, you receive various warnings and error messages in the SAS log and, if the step continues to execute, you may get unpredictable results such as errors in character values. In addition, when running the SAS System under display manager, mismatched or missing quotes can cause more serious errors.**

Once you have checked your program and are satisfied that it contains no syntax errors, you can submit it for processing.

---

*   The SAS System processes some statements containing misspelled keywords. For example, if you misspell the keywords DATA or PROC, the system attempts to interpret what you meant, continues processing your SAS program based on its interpretation, and writes a message to the SAS log telling you how it interpreted what you typed. Other misspellings, such as misspelled variable names, generate error messages and may prevent complete processing.

**   For example, the SAS System will not finish a step that contains missing or unmatched quotes until it processes a matching quote. You will see the message "DATA step running" or "PROC step running" in the upper right corner of the PROGRAM EDITOR window. To solve this problem, from the PROGRAM EDITOR window type a quote followed by a semicolon and a RUN statement followed by a semicolon, and submit these statements. The step should execute. Once the step executes, go back and correct the original program.

Understanding SAS Log Messages

When the SAS System processes programs, it generates a variety of information, which is displayed in the SAS log. The SAS log displays the SAS statements you submit for processing and contains system messages about the execution of your program. When the system detects an error, it underlines the error or the point at which it detects the error and identifies the error by number. The system then responds by issuing one of three types of messages: notes, warnings, or error messages. The boldface numbers and shaded lines in Output 12.1 correspond to the following items:

1. A *note* is an informative message or explanation. For example, the first note in Output 12.1 informs you of the number of observations and variables in the newly created data set. The second note tells you that processing stopped due to errors.

2. A *warning* informs you of a potential problem. For example, in Output 12.1 the keyword DATA is misspelled as DATTA. The SAS System interprets the word DATTA as a misspelling of DATA and continues processing the program. At the end of the DATA step, the SAS System writes a warning to the SAS log informing you of its action.

3. A SAS *error message* informs you when the system encounters an error, such as when a statement is invalid or used out of order. Output 12.1 contains an error message indicating that an equal sign (=) is missing before the data set name in the PROC PRINT statement.

Output 12.1
Log Messages

```
5          datta htwt;
           14
6              input name $ 1-10 sex $ 12 age 14-15 height 17-18 weight 20-22;
7              cards;

WARNING 14-169: Assuming the symbol DATA was misspelled as DATTA.  2

NOTE: The data set WORK.HTWT has 18 observations and 5 variables.  1

26         ;
27             proc print data htwt;
                            73
28             title 'Heights, Weights and Ages';
29         run;

ERROR 73-322: Expecting an =.   3

NOTE: The SAS System stopped processing this step because of errors.  1
```

The numbers next to warnings and error messages are part of an internal numbering system that the SAS System uses to locate the text of messages.

Verifying Data

If the SAS System produces output that appears incorrect, you can trace the problem in some cases by checking your data. For example, check the log note that tells you the number of observations and variables written to a data set to make sure the DATA step read and processed all the raw data lines you input.

The LIST statement causes the SAS System to write to the log the input data records for the observation being processed. This is a useful method for discovering errors in raw data lines. In Output 12.2, the SAS System lists the input data record when the value of AGE is missing.

Output 12.2
LIST Statement

```
5          data htwt;
6              input name $ 1-10 sex $ 12 age 14-15 height 17-18 weight 20-22;
7              if age=. then list;
8              cards;

RULE:      ----+----1----+----2----+----3----+----4----+----5----+----6----+----
23         Yao        M  .  70 145
NOTE: The data set WORK.HTWT has 18 observations and 5 variables.

27         run;
```

Note: The LIST statement only works with raw data lines read with the INPUT statement; it does not write observations read from other SAS data sets with the SET statement.

It is sometimes useful to produce a list in the SAS log of variable names and values for an observation. You can do this with the PUT _ALL_ statement.

The PUT _ALL_ statement can be used throughout SAS programs as a way to check programming logic. For example, you know that the observation that has a value of **Yao** for the NAME variable has a missing value for the AGE variable. You can check the existing value, provide a new value, and then quickly verify that the correction was made by using IF/THEN, assignment, and PUT _ALL_ statements. Note that the DATA step creates a data set called NEW using the observations from the data set HTWT read with the SET statement. Output 12.3 shows the results.

Output 12.3
*PUT _ALL_*
Statement

```
27         data new;
28             set htwt;
29             if name='Yao' then put _all_;
30             if name='Yao' then age=53;
31             if name='Yao' then put _all_;
32         run;

NAME=Yao SEX=M AGE=. HEIGHT=70 WEIGHT=145 _ERROR_=0 _N_=15
NAME=Yao SEX=M AGE=53 HEIGHT=70 WEIGHT=145 _ERROR_=0 _N_=15
NOTE: The data set WORK.NEW has 18 observations and 5 variables.
```

Note: The SAS System creates two automatic variables, _ERROR_ and _N_, that are assigned temporarily to each observation. These variables are not stored with the data set.

For more information on understanding errors, see Chapter 23, "Diagnosing and Avoiding Errors," in *SAS Language and Procedures: Usage, Version 6, First Edition* and Chapter 5, "SAS Output," in *SAS Language: Reference, Version 6, First Edition.*

Appendix 1 Using the SAS® Display Manager System

Introduction

The *SAS Display Manager System* is an interactive full-screen facility that you view and operate through a series of windows. This appendix presents basic information about the SAS Display Manager System. It explains how to use display manager to edit text, submit SAS programs, manage program output, and understand the results of your program.

The SAS Display Manager System: Learning the Basics

The SAS Display Manager System is an interactive full-screen facility that you view and operate through a series of windows. As an interactive facility, it enables you to accomplish a series of tasks rather than one task; as a full-screen facility, it enables you to view, alter, and submit more than one line of SAS statements at a time. It is therefore considered a convenient way to interact with the SAS System.

Because display manager is composed of windows, it is often called the SAS windowing environment. Within this windowing environment, you issue commands to perform such tasks as retrieving, editing, submitting, and resubmitting all or part of a SAS program within the same SAS session. In addition, many SAS display manager windows enable you to perform specific interactive applications, such as changing the settings of function keys. These windows are discussed fully in Chapter 17, "SAS Display Manager Windows," in *SAS Language: Reference, Version 6, First Edition.*

The following sections describe display manager and explain how to employ some of the most frequently used windows and commands to process data.

Display Manager Windows

The SAS Display Manager System consists of four primary windows: the PROGRAM EDITOR window, LOG window, OUTPUT window, and OUTPUT MANAGER window. Each window enables you to perform specific types of tasks. The following sections describe the primary windows in detail.

Invoking Windows

The PROGRAM EDITOR (PGM) and LOG windows are the first windows that appear when you invoke display manager. Both are open, and the PROGRAM EDITOR window contains the cursor and is therefore the active window. An *active window* is open and displayed and contains the cursor. Only one window can be active at a time. You can open display manager windows using window-call commands, usually the name or nickname of the window you want to open. When a window is invoked with a window-call command, it is open, displayed, and active. Windows are typically closed with the END command.

Depending on how the SAS System is installed at your location, the windowing environment may look different from the displays shown in this appendix. At some sites, the default for running display manger is the PMENU facility, a menuing system that replaces the command line as a way to issue commands. The PMENU facility is discussed in more detail later in this appendix.

Entering Text and Understanding Program Results

The PROGRAM EDITOR window enables you to enter, edit, copy, and store text, and to submit SAS programs. The PROGRAM EDITOR window can be used to issue most frequently used display manager commands and to access other windows. Because of its many capabilities, you may use this window most often.

The LOG window contains the SAS log. As you may recall from previous chapters, the SAS log contains notes, warnings, error messages, and other SAS System information that enables you to identify and correct program errors. The PROGRAM EDITOR and LOG windows are illustrated in Display A1.1.

Display A1.1
PROGRAM
EDITOR and LOG
Windows

```
┌LOG────────────────────────────────────────────────────────────────────────
│ Command ===>
│
│ 4    data htwt;
│ 5        input name $ 1-10 sex $ 12 age 14-15 height 17-18 weight 20-22;
│ 6        cards;
│
│ NOTE: The data set WORK.HTWT has 18 observations and 5 variables.
│
│ 25   ;
│ 26   run;
│
│
│
│
│
│
│
└────────────────────────────────────────────────────────────────────────────
```

```
┌PROGRAM EDITOR───────────────────────────────────────────────────────────────
│ Command ===>
│
│ 00001 proc print data=htwt;
│ 00002 run;
│ 00003 proc plot data=htwt;
│ 00004     plot height*weight;
│ 00005 run;
│ 00006
│ 00007
│ 00008
│ 00009
└────────────────────────────────────────────────────────────────────────────
```

The windows in Display A1.1 have characteristics that are shared by most other display manager windows. These include a command line located in the upper left corner of the display (when you are not using the PMENU facility), a border around the window, and the name of the window in the upper left corner of the top of the window border.

Managing Program Output

The OUTPUT window contains output from procedures. If you submit the programs in this book using display manager, output shown in the examples goes to the OUTPUT window. To view the results of a program in the OUTPUT window, specify LISTING ON. You can browse and scroll procedure output there. Display A1.2 shows the OUTPUT window with output produced by the PRINT procedure.

Display A1.2
OUTPUT Window

```
┌OUTPUT─────────────────────────────────────────────────────────────┐
│ Command ===>                                                       │
│ NOTE: Procedure PRINT created 1 page(s) of output.                 │
│                      Height-Weight Study                         1 │
│                                                                    │
│          OBS     NAME         SEX    AGE    HEIGHT    WEIGHT        │
│                                                                    │
│            1     Aubrey        M      41      74        170         │
│            2     Ron           M      42      68        166         │
│            3     Carl          M      32      70        155         │
│            4     Antonio       M      39      72        167         │
│            5     Deborah       F      30      66        124         │
│            6     Jacqueline    F      33      66        115         │
│            7     Helen         F      26      64        121         │
│            8     David         M      30      71        158         │
│            9     James         M      53      72        175         │
│           10     Michael       M      32      69        143         │
│           11     Ruth          F      47      69        139         │
│           12     Joel          M      34      72        163         │
│           13     Donna         F      23      62         98         │
│           14     Roger         M      36      75        160         │
│           15     Yao           M       .      70        145         │
│           16     Elizabeth     F      31      67        135         │
│           17     Tim           M      29      71        176         │
│           18     Susan         F      28      65        131         │
│                                                                    │
│                                                                    │
└────────────────────────────────────────────────────────────────────┘
```

The OUTPUT MANAGER window acts as an index for your OUTPUT window. To view the contents of the OUTPUT MANAGER window, specify MANAGER ON. As you can see in Display A1.3, the window provides the procedure name, page number and length, and description based on the first 40 characters of the title.

Display A1.3
OUTPUT
MANAGER
Window

```
┌OUTPUT MANAGER──────────────────────────────────────────────────────┐
│ Command ===>                                                       │
│                                                                    │
│      Procedure   Page#    Pages        Description                 │
│    _ PRINT         1        1           Height-Weight Study         │
│    _ PLOT          2        1           Plot of Height*Weight       │
│    _ PRINT         3        1           Names and Ages of Study Participants │
│                                                                    │
│                                                                    │
│                                                                    │
│                                                                    │
│                                                                    │
│                                                                    │
│                                                                    │
│                                                                    │
│                                                                    │
│                                                                    │
│                                                                    │
│                                                                    │
└────────────────────────────────────────────────────────────────────┘
```

You can also use the OUTPUT MANAGER window to edit, print, save, and delete output, in addition to other tasks, such as renaming the output's description. For more information about invoking the OUTPUT and OUTPUT MANAGER windows and toggling between the two, see "OUTPUT and OUTPUT MANAGER Windows" in Chapter 7, "SAS Display Manager System," in *SAS Language: Reference.*

Supplying Information to Requestor Windows

A *requestor window* is a window that the SAS System displays to prompt you to choose a course of action or provide you with information. For example, suppose you issue a FILE command to write a program to an external file. If the file you name already exists, a requestor window prompts you to indicate whether you want to replace or append the contents of the file or cancel the operation. Display A1.4 shows the requestor window that appears in this example.

Display A1.4
Requestor Window

```
┌LOG─────────────────────────────────────────────────────────────────┐
│ Command ===>                                                        │
│                                                                     │
│                                                                     │
│                                                                     │
│                                                                     │
│                                                                     │
│                                                                     │
│         --------------------------------------------------------    │
│         |Warning:  The file already exists.  Enter R to replace it, |│
│         |enter A to append to it or C to cancel FILE command.     | │
│         |_                                                        | │
│         --------------------------------------------------------    │
└─────────────────────────────────────────────────────────────────────┘
┌PROGRAM EDITOR────────────────────────────────────────────────────────┐
│ Command ===>                                                         │
│                                                                      │
│ 00001 data htwt;                                                     │
│ 00002     input name $ 1-10 sex $ 12 age 14-15 height 17-18 weight 20-22; │
│ 00003     cards;                                                     │
│ 00004 Aubrey      M 41 74 170                                        │
│ 00005 Ron         M 42 68 166                                        │
│ 00006 Carl        M 32 70 155                                        │
│ 00007 Antonio     M 39 72 167                                        │
│ 00008 Deborah     F 30 66 124                                        │
│ 00009 Jacqueline  F 33 66 115                                        │
└──────────────────────────────────────────────────────────────────────┘
```

To exit a requestor window, you must provide the information requested.

Display manager contains additional windows used to manage SAS data sets and libraries, change function key settings, and obtain online help, among other tasks. Some of these windows are detailed later in this appendix.

In the next section, you will learn what you can do with display manager commands.

Display Manager Commands

Display manager commands enable you to display a window, alter its configuration, edit text, and globally search for and alter selected words or phrases. These commands are divided into two categories: basic commands and editing commands.

Using Basic Commands

Basic commands are those display manager commands that perform basic tasks such as scrolling (moving the text vertically or horizontally across the display), managing files, and managing windows. You can issue basic commands in one of three ways, depending on how the SAS System runs at your site:

□ from the command line (when running display manager without the PMENU facility)

□ through the PMENU facility

□ with function keys.

Basic commands issued from the command line are referred to in other base SAS software documentation as command-line commands. They are issued by typing the command and pressing the ENTER or RETURN key, depending on the host operating system you are using. The following example shows how to issue the SUBMIT command from the command line:

```
Command===> submit
```

The *PMENU facility* is a menuing system through which you issue display manager commands. At some sites, this facility is the default setting. If not the default but if available at your site, you can invoke the PMENU facility by issuing the PMENU command.

The PMENU facility contains a menu, or *action bar*, of several items that you can select by placing the cursor on the item and pressing ENTER or RETURN. The action bar replaces the command line when you are using the PMENU facility. Once you select a menu item, the SAS System executes the command, displays a *pull-down menu* containing additional selections, or displays a dialog box that prompts you for information. A *dialog box* is similar to a requestor window in that it appears as a response to something you do or attempt to do. Unlike requestor windows, you can exit a dialog box without supplying the information requested by issuing the CANCEL command.

Display A1.5 gives an example of a PMENU display, showing a dialog box and a pull-down menu generated from selecting items on the action bar in the PROGRAM EDITOR window. Note that the LOG window also contains an action bar.

Display A1.5
PMENU Facility

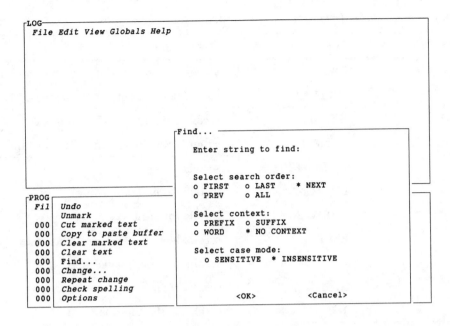

To issue a basic command using a function key, press the key to which the command is assigned. Display manager automatically executes the command. You can browse, alter, and save function key settings with the display manager KEYS window, discussed later in this appendix.

Basic commands and descriptions

Some frequently used commands are those which enable you to maneuver the cursor on the display. For example, the FORWARD command scrolls the text in the display forward, and the BACKWARD command scrolls back. The TOP command takes the cursor to the top of the file, and the BOTTOM command takes it to the end of the file.

Other commands enable you to open and display a window, making it the active window. These commands correspond to the name or nickname of each window. For example, the HELP command opens the HELP window, and the KEYS command opens the KEYS window.

Once you have opened a window, you can manage windows with display manager commands. The CLEAR command, for example, removes the contents of an active window. The ZOOM command increases the window size, and the BYE command ends a SAS session. The END command closes a window and removes it from the display.

You can search for and change text strings in most windows using display manager commands. For example, the FIND command searches for a text string; the RFIND command repeats the search without requiring you to respecify the text string. The CHANGE and RCHANGE commands work similarly, changing a text string and repeating the change for the next occurrence of the string.

These commands are just a few of the most commonly used basic commands in display manager. Table A1.1 contains an expanded list of additional commands, the category each command falls in based on the type of action it performs, and a description of the action. For a complete table of display manager commands, see Chapter 7, "SAS Display Manager System," in *SAS Language: Reference*.

Table A1.1   *Basic Commands and Descriptions*

Category	Command	Description
file-management	FILE	writes the contents of a window to an external file
	INCLUDE	copies the contents of an external file into a window
program	RECALL	recalls submitted SAS statements
	SUBMIT	submits SAS statements for execution
scrolling	BACKWARD	scrolls backward
	BOTTOM	scrolls to the bottom line
	FORWARD	scrolls forward
	LEFT	scrolls left
	n	scrolls to a designated line
	RIGHT	scrolls right
	TOP	scrolls to the top line
search	BFIND	searches for the previous occurrence of a character string
	CHANGE	finds and changes one character string to another
	FIND	searches for a specified character string
	RCHANGE	repeats the previous CHANGE command
	RFIND	continues the search initiated with a FIND or BFIND command
text-edit	CURSOR	moves the cursor to the command line
	RESET	removes any pending line commands
	UNDO	reverses the effects of some actions
window-call	AF	invokes the AF window
	APPOINTMENT	invokes the APPOINTMENT window
	CALCULATOR	invokes the CALCULATOR window
	CATALOG	invokes the CATALOG window
	DIR	invokes the DIR window
	FILENAME	invokes the FILENAME window
	FOOTNOTES	invokes the FOOTNOTES window
	FSFORM	invokes the FORM window
	HELP	invokes the HELP window
	KEYS	invokes the KEYS window
	LIBNAME	invokes the LIBNAME window

(continued)

Table A1.1 (continued)

Category	Command	Description
window-call	LISTING	invokes the OUTPUT window
	LOG	invokes the LOG window
	MANAGER	invokes the OUTPUT MANAGER window
	NOTEPAD	invokes the NOTEPAD window
	OPTIONS	invokes the OPTIONS window
	OUTPUT	invokes either the OUTPUT window or the OUTPUT MANAGER window
	PROGRAM	invokes the PROGRAM EDITOR window
	SETINIT	invokes the SETINIT window
	SITEINFO	invokes the SITEINFO window
	TITLES	invokes the TITLES window
	VAR	invokes the VAR window
window-management	BYE	ends a SAS session
	CANCEL	cancels changes in a window and removes it from the display
	CLEAR	clears the window's contents permanently or the display of settings
	END	closes a window and removes it from the display
	ENDSAS	ends a SAS session
	HOME	moves the cursor to the command line
	ICON	makes the active window a smaller version of itself
	NEXT	moves the cursor to the next window, activating it
	PMENU	activates or deactivates the PMENU facility for all windows
	PREVCMD	recalls the last command issued
	PREVWIND	moves the cursor to the previous window, activating it
	ZOOM	causes the active window to fill the display

Using Editing Commands

Editing commands copy, move, delete, and otherwise edit text. They can also be used to justify lines of text, insert blank lines, or shift lines of text to the left or right.

Editing commands can be grouped into two categories. In the first category are commands used for cutting, pasting, and storing text and issued from the command line, with function keys, or using the PMENU facility. These commands are the CUT, MARK, PASTE, PCLEAR, PLIST, SMARK, STORE, and UNMARK commands. They are described in Chapter 18, "SAS Display Manager Commands," in *SAS Language: Reference.*

In the second category are line commands. These are editing commands that are usually entered from the numbered field in the left-most portion of the

PROGRAM EDITOR window. This section describes how to edit text using line commands.

Line commands can be entered anywhere within the numbered field in the left-most portion of the PROGRAM EDITOR window. Line commands can also be assigned to function keys and then executed by pressing the function key. A less common way of issuing line commands is from the command line, by typing a colon (:) followed by the line command.

Editing commands and descriptions

Editing commands are frequently used to move or copy text in the PROGRAM EDITOR window. For example, you can use the M (move) command to make the following list alphabetical. Type the letter M anywhere in the numbered field next to the text you want to move. Specify a destination by typing the letter A (after) or B (before) on the line you want the moved text to follow or precede, and press ENTER or RETURN. Display A1.6 shows the text after you have typed the line commands but before you have pressed ENTER or RETURN.

Display A1.6
Move Operation

```
┌PROGRAM EDITOR──────────────────────────────────────────────────────┐
│  Command ===>                                                        │
│                                                                      │
│ m0001 Yao                                                            │
│ 00002 Donna                                                          │
│ 00003 Helen                                                          │
│ 00004 James                                                          │
│ 0a005 Roger                                                          │
│ 00006                                                                │
│ 00007                                                                │
│ 00008                                                                │
│ 00009                                                                │
│                                                                      │
└──────────────────────────────────────────────────────────────────────┘
```

After you press the ENTER or RETURN key, the results appear as shown in Display A1.7.

Display A1.7
Results of Move
Operation

```
┌PROGRAM EDITOR──────────────────────────────────────────────────────┐
│  Command ===>                                                        │
│                                                                      │
│ 00001 Donna                                                          │
│ 00002 Helen                                                          │
│ 00003 James                                                          │
│ 00004 Roger                                                          │
│ 00005 Yao                                                            │
│ 00006                                                                │
│ 00007                                                                │
│ 00008                                                                │
│ 00009                                                                │
│                                                                      │
└──────────────────────────────────────────────────────────────────────┘
```

The ground rules are similar for working with a block of text. Type double letters anywhere in the numbered area of the beginning and ending lines you want to edit. For example, to alphabetize the following list, you must move a block of text. Note the MM (move) block command on lines 4 and 5 and the B line command on line 1 in Display A1.8.

Display A1.8
Block Move
Operation

```
┌PROGRAM EDITOR─────────────────────────────────────────────────────┐
│ Command ===>                                                      │
│                                                                   │
│ b0001 James                                                       │
│ 00002 Roger                                                       │
│ 00003 Yao                                                         │
│ mm004 Donna                                                       │
│ mm005 Helen                                                       │
│ 00006                                                             │
│ 00007                                                             │
│ 00008                                                             │
│ 00009                                                             │
│                                                                   │
└───────────────────────────────────────────────────────────────────┘
```

The operation is complete after you press the ENTER or RETURN key. Display A1.9 contains the results.

Display A1.9
Results of Block
Move Operation

```
┌PROGRAM EDITOR─────────────────────────────────────────────────────┐
│ Command ===>                                                      │
│                                                                   │
│ 00001 Donna                                                       │
│ 00002 Helen                                                       │
│ 00003 James                                                       │
│ 00004 Roger                                                       │
│ 00005 Yao                                                         │
│ 00006                                                             │
│ 00007                                                             │
│ 00008                                                             │
│ 00009                                                             │
│                                                                   │
└───────────────────────────────────────────────────────────────────┘
```

These principles apply to the C (copy) and D (delete) line commands as well.

Table A1.2 lists some common display manager editing commands and their descriptions.

Table A1.2
Editing Commands
and Descriptions

Command	Description
C and CC	copy one or more lines before (B) or after (A) the target line
CL and CCL	lowercase all characters in one or more designated lines of text
COLS	displays a line ruler that marks horizontal columns
CU and CCU	uppercase all characters in one or more designated lines of text
D and DD	delete one or more lines
I	inserts one or more new lines
M and MM	move one or more lines of text before (B) or after (A) the target line
R and RR	repeat one or more designated lines of text
TC	connects two lines of text
TF	flows text to a blank line or to the end of text
TS	splits text at the cursor
> and >>	shift right one or more designated lines of text
< and <<	shift left one or more designated lines of text

Line commands are documented elsewhere as part of the SAS Text Editor; however, it is beyond the scope of this book to fully discuss the text editor. For complete reference information on display manager and text editor commands, refer to Chapter 7; Chapter 8, "SAS Text Editor;" Chapter 18, "SAS Display Manager Commands;" and Chapter 19, "SAS Text Editor Commands;" in *SAS Language: Reference.*

The SAS Display Manager System: Learning More

In addition to its four primary windows, display manager contains many other windows that enable you to control the windowing environment. The next sections describe the HELP, KEYS, FILENAME, and LIBNAME windows in more detail. For complete reference information on all display manager windows, see Chapter 17 in *SAS Language: Reference.*

Getting Additional Information Using the HELP Window

As you use the SAS Display Manager System, you may find occasionally that you need help. Perhaps you're familiar with a command or a window but don't fully understand what it accomplishes. Or perhaps you know what you want to do but don't know which command or window to use. In either case, you can obtain the information you need online, from the HELP window. To access the HELP window, issue the HELP command.

When accessed from the PROGRAM EDITOR window, the HELP window appears as shown in Display A1.10.

Display A1.10
HELP Window

```
┌HELP: SAS System Help─────────────────────────────────────────────
│ Command ===>
│
│     SAS SYSTEM HELP: Main Menu
│
│
│     DATA MANAGEMENT           REPORT WRITING            GRAPHICS
│
│
│     TUTORIAL           MODELING & ANALYSIS TOOLS       UTILITIES
│
│
│     SAS LANGUAGE           SAS GLOBAL COMMANDS        SAS WINDOWS
│
│
│                       HOST INFORMATION
│
│
│                            INDEX
│
│
│
├──────────────────────────────────────────────────────────────────
│ 00003
│ 00004
│ 00005
│ 00006
│ 00007
│ 00008
│ 00009
```

Note: Depending on your terminal, the HELP window may look slightly different from this display and may cover all or part of the display.

From the SAS System Help Main Menu, you can request help information for base SAS software procedures, windows, other components of base SAS software, and other SAS software products. The Main Menu also includes an index. Move the cursor to the category you want and press ENTER or RETURN, or use a mouse to point and click. If you issue the HELP command within SAS System Help, you see information on how to use SAS System Help.

From any help window, issuing the END command closes the current help window and removes it from the display, returning to the previous window. You can exit the help system directly from any window in SAS System Help, bypassing previous help windows, by issuing the =X command.

For help on display manager commands discussed earlier in this chapter, select SAS GLOBAL COMMANDS from the Main Menu. For information on specific command-line and line commands, select SAS Text Editor from the SAS Global Commands window.

As you become more familiar with the help windows, you can take shortcuts to obtain the information you want. From any of the primary windows in display manager, you can move directly to a help window for the part of the SAS System you choose.

For example, to access a help window for the PRINT procedure, specify the HELP command followed by PRINT, as follows:

```
COMMAND===> help print
```

The SAS System bypasses the Main Menu and displays a help window containing the index entry for the PRINT procedure. From there, you can obtain more information about the PRINT procedure by selecting Introduction or Syntax, or you can return to the previous display manager window by issuing the END command.

Simplifying Commands Using the KEYS Window

Most terminals have special keys called function keys. With them, you can use one keystroke to issue a command. Depending on the type of terminal you're using, the SAS System assigns default definitions to your terminal's function keys. The KEYS window shows you what commands are assigned to keys and enables you to set up your own key definitions.

To access the KEYS window, issue the KEYS command. A sample of typical key settings is displayed in the KEYS window in Display A1.11.

Display A1.11
KEYS Window

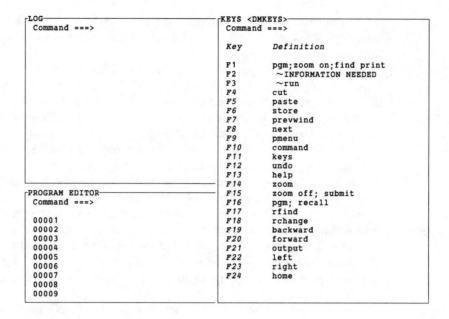

```
┌LOG─────────────────────────────┐  ┌KEYS <DMKEYS>──────────────────┐
│  Command ===>                   │  │  Command ===>                  │
│                                 │  │                                │
│                                 │  │    Key      Definition         │
│                                 │  │                                │
│                                 │  │    F1       pgm;zoom on;find print │
│                                 │  │    F2       ~INFORMATION NEEDED │
│                                 │  │    F3       ~run               │
│                                 │  │    F4       cut                │
│                                 │  │    F5       paste              │
│                                 │  │    F6       store              │
│                                 │  │    F7       prevwind           │
│                                 │  │    F8       next               │
│                                 │  │    F9       pmenu              │
│                                 │  │    F10      command            │
│                                 │  │    F11      keys               │
│                                 │  │    F12      undo               │
│                                 │  │    F13      help               │
│                                 │  │    F14      zoom               │
├PROGRAM EDITOR──────────────────┤  │    F15      zoom off; submit   │
│  Command ===>                   │  │    F16      pgm; recall        │
│                                 │  │    F17      rfind              │
│  00001                          │  │    F18      rchange            │
│  00002                          │  │    F19      backward           │
│  00003                          │  │    F20      forward            │
│  00004                          │  │    F21      output             │
│  00005                          │  │    F22      left               │
│  00006                          │  │    F23      right              │
│  00007                          │  │    F24      home               │
│  00008                          │  │                                │
│  00009                          │  │                                │
└─────────────────────────────────┘  └────────────────────────────────┘
```

Scroll forward to view all of the default settings. To change the definition of any setting, type the new definition over an old one. Once defined, a setting takes effect immediately, whether or not you exit the KEYS window. You can change your key settings to any of the display manager commands already discussed in this appendix.

To assign a series of commands to a function key, separate them with semicolons. For example, the F16 key in Display A1.11 has two commands assigned to it. Pressing the F16 key moves the cursor to the PROGRAM EDITOR window (the PGM command) and then displays previously submitted program statements or text (the RECALL command). Notice that many of the display manager commands described earlier in this appendix are assigned to the function keys in Display A1.11.

If you have changed your settings but want to cancel the change before exiting the KEYS window, issue the CANCEL command. Your previous settings are restored. To exit the KEYS window, issue the END command.

Aside from changing or adding commands within the KEYS window, you can also define a key to insert text into the PROGRAM EDITOR window. Type a tilde (~) in the first column of the field followed by the text you want to insert. Then every time you press that function key, the text string (without the tilde symbol) is inserted at the location of the cursor. In Display A1.11, the F2 and F3 keys are defined to include text using this method. Depending on the operating system you're using, the inserted text may or may not overlay text following the cursor.

Exploring SAS Data Sets Using the LIBNAME, DIR, and VAR Windows

A *SAS data library* is a collection of SAS files, including SAS data sets. You can use a LIBNAME statement to associate a libref, or library reference name, with a SAS data library. Then as you write SAS programs and submit SAS statements in display manager, you can use the libref as a shorthand way to refer to the library.

Issue the LIBNAME command to display a list of current librefs and the SAS data library names they reference in the LIBNAME window. Display A1.12 shows the LIBNAME window.

Display A1.12
LIBNAME Window

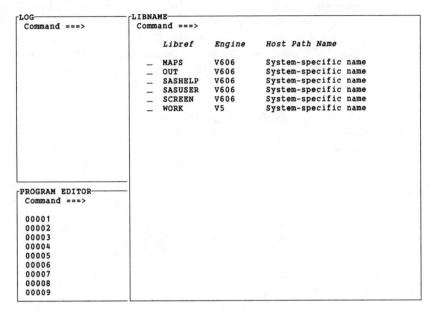

The LIBNAME window contains the libref, the name of the engine used to access the library, and the name of the corresponding SAS data library.* The phrase System-specific name represents the name of a file from your host operating system. This window is useful if you are working with either temporary or permanent SAS data sets as discussed in "Special Topic: Temporary and Permanent SAS Data Sets" in Chapter 3, "Creating SAS Data Sets." For example, the HTWT data set created in the beginning of this appendix is a temporary SAS data set that resides in the data library referenced by the WORK libref.

To list the SAS data sets contained in one of the SAS data libraries, move the cursor to the selection field to the left of the libref you choose, type X, and press ENTER or RETURN. The DIR window appears, listing all the SAS data sets in that SAS data library.

From the DIR window, you can browse a list of a data set's observations, rename or delete a file, or access other display manager windows for more information about the file. To access the VAR window for a file, move the cursor to the selection field next to the file, type X, and press ENTER or RETURN. The VAR window displays information about variables and their attributes for the SAS data set specified, providing a quick overview of the data set's contents.

---

* It is beyond the scope of this book to discuss an advanced topic such as SAS engines. For additional information, refer to *SAS Language: Reference* or *SAS Language and Procedures: Usage, Version 6, First Edition.*

Accessing External Files Using the FILENAME Window

Although you may store much of your data in SAS files, you may also store your data in external files. You can use a FILENAME statement to associate a fileref, or file reference name, with an external file. Then, as you write SAS programs and submit SAS statements in display manager, you can use the fileref as a shorthand way to refer to the external file. Issue the FILENAME command to display a list of current filerefs and the names of the files they reference in the FILENAME window. The fileref is shown in the left-hand column, and the file's complete name appears in the right-hand column, as shown in Display A1.13. Note that System-specific name represents the complete name of the external file on your operating system.

Display A1.13
FILENAME
Window

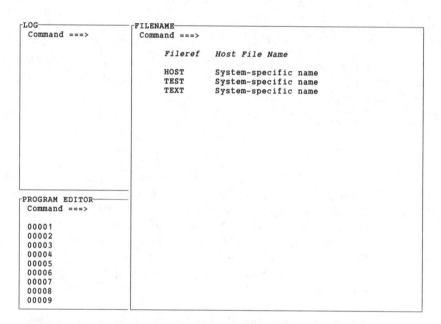

You can also use many of the display manager commands already discussed in this chapter in the FILENAME window. As with the DIR and LIBNAME windows, you can exit the FILENAME window by issuing the END command.

For more information on using display manager windows, refer to Chapter 7 and Chapter 17 in *SAS Language: Reference*.

Glossary

action bar

a list of selections that appear when the PMENU command is executed. The action bar is used by placing your cursor on the selection that you want and pressing ENTER. This either executes a command or displays a pull-down menu.

active window

a window that is open, displayed, and contains the cursor. Only one window can be active at a time.

arithmetic operator

a symbol used to perform arithmetic calculations, such as addition, subtraction, multiplication, division, and exponentiation.

assignment statements

statements that enable you to create variables and modify existing variable values. Assignment statements assign values to specified variables.

batch mode

on mainframes and minicomputers, a method of running SAS programs in which you prepare a file containing SAS statements and any necessary operating system control statements and submit the file to the operating system. Execution is completely separate from other operations at your terminal and is sometimes referred to as running in background.

BY variable

a variable named in a BY statement.

cell

a single unit of a table produced by a SAS procedure, such as the FREQ procedure. The value contained in the cell is a summary statistic for the input data set. The contents of the cell are described by the page, row, and column that contain the cell.

character variable

a value that can contain alphabetic characters, numeric characters 0 through 9, and other special characters. Character variables contain character values.

chart variable

the variable whose values you are charting.

column input

a style that gives column specifications in the INPUT statement for reading data entered in fixed columns.

column percent

the percentage of observations represented in a column of output.

comparison operator
a symbol or two-character abbreviation used to test a relationship between two values, such as > (GT).

crosstabulation table
a frequency table that displays the frequency distribution for two or more variables. These tables are often referred to as two-way, three-way, or n-way tables. See also frequency table.

cumulative frequency
the number of observations in all ranges up to and including a given range.

cumulative percent
the percentage of observations in all ranges up to and including a given range.

data error
a type of execution error that occurs when the data being analyzed by a SAS program contain invalid values. For example, a data error occurs if you specify numeric variables in the INPUT statement for character data. Data errors do not cause a program to stop, but instead they produce notes.

data lines
lines of unprocessed data, often referred to as raw data.

DATA step
a group of statements in a SAS program that begin with a DATA statement and end with a RUN statement, another DATA statement, a PROC statement, or the end of the job. The DATA step enables you to read raw data or other SAS data sets and to create SAS data sets.

data value
(1) a single unit of information, such as one person's height. (2) the intersection of a row (observation) and column (variable) in the rectangular form of a SAS data set.

dialog box
a feature of the PMENU facility that appears in response to an action, usually selecting a menu item. The purpose of dialog boxes is to obtain information, which you supply by filling in a field or choosing a selection from a group of fields. You can execute the CANCEL command to exit the dialog box.

display manager mode
an interactive windowing method of running SAS programs in which you edit a group of statements, submit the statements, and then review the results of the statements in various windows.

error message
a message that informs you when the SAS System encounters an error, such as when a statement is invalid or used out of order.

execution-time error
error detected when the system executes a step.

external file
a file created and maintained on the host operating system from which you can read data or store SAS statements or in which you can store procedure output. An external file is not a SAS data set.

fileref
the name used to identify an external file to the SAS System. You assign a fileref with a FILENAME statement.

frequency chart
a graphic illustration of the number of times a value or range of values for a given variable occurs in a data set.

frequency count
the number of times a value or range of values for a given variable occurs in a data set.

frequency table
a table that summarizes data by displaying frequency counts. Frequency tables that process one variable are often referred to as one-way tables. See also frequency count and crosstabulation table.

global statements
statements that remain in effect until you cancel them.

group variable
a variable used to define separate sets of bars or blocks in a single chart.

interactive line mode
a method of running SAS programs without using the SAS Display Manager System. You enter one line of a SAS program at a time. The SAS System processes each line immediately after you enter it.

libref
the name temporarily associated with a SAS data library.

list input
a style that supplies only variable names, not column locations, in the INPUT statement, enabling you to enter data values separated by at least one blank.

logical operator
an operator used in expressions to link sequences of comparisons. The logical operators are AND, OR, and NOT.

methods of running the SAS System
one of the following modes used to run SAS programs: display manager mode, interactive line mode, noninteractive mode, batch mode.

midpoints
values that identify the bars, blocks, or sections of a chart. A midpoint value represents a range of values or a single value.

missing values

values that represent missing or unavailable data values to the SAS system. You represent missing values with periods or blanks, depending on the method of data entry and type of data value. The SAS System displays a blank to represent a missing value for a character variable and a period or a special character to represent a missing value for a numeric variable.

noninteractive mode

a method of running SAS programs in which you prepare a file of SAS statements and submit the program to the computer system. The program runs immediately and occupies your current terminal session.

nonstandard data

data that appear in alternate or nonstandard forms, such as numbers containing commas.

note

in SAS logs, an informative message or explanation.

numeric variable

a variable that contains only numeric values and related symbols, such as decimal points, plus signs, and minus signs.

observation

a set of data values for the same entity, for example all physical measurements for one person.

permanent SAS data set

a data set that remains after the current program or interactive SAS session terminates. Data stored in permanent SAS data sets can be retrieved for use in future programs or sessions.

PMENU facility

a menuing system that is used instead of the command line as a way to execute commands.

PROC step

a group of SAS statements that call and execute a procedure, usually with a SAS data set as input.

procedures

a collection of built-in SAS programs that are used to produce reports, manage files, and analyze data. They enable you to accept default output or to tailor your output by overriding defaults.

pull-down menu

the list of choices that appear when you choose an item from an action bar or from another pull-down menu in the PMENU facility. The choices in the list are called items.

raw data

data that have not been read into a SAS data set. See also data lines.

requestor window

a window that the SAS System displays so that you can confirm, cancel, or modify an action.

row percent

the percentage of observations represented in a row of output.

SAS data library

a collection of one or more SAS files, including SAS data sets, that are recognized by the SAS System. Each data set is a member of the SAS data library.

SAS data set

data values that are organized as a table of observations and variables that can be processed by the SAS System.

SAS Display Manager System

an interactive windowing environment in which actions are performed with a series of commands or function keys. Within one session, multiple tasks can be accomplished. It can be used to prepare and submit programs, view and print the results, and debug and resubmit the programs.

SAS keyword

a literal that is a primary part of the SAS language. Common SAS keywords include DATA, INFILE, INPUT, PROC, and other statement names.

SAS log

a file that can contain the SAS statements you enter and messages about the execution of your program.

SAS procedures

See procedures.

SAS program

a sequence of related SAS statements.

SAS statement

a string of SAS keywords, SAS names, and special characters and operators ending in a semicolon that instruct the SAS System to perform an operation or give information to the SAS System.

subgroup variable

a variable that segments the bars or blocks of a chart. The CHART procedure fills each bar or block with characters that show the contribution of each value of the subgroup variable to the total.

syntax error

an error in the spelling or grammar of SAS statements. The SAS System finds syntax errors as it compiles each SAS step before execution.

temporary SAS data set

a data set that exists only for the duration of the current program or interactive SAS session. Therefore, data stored in temporary SAS data sets cannot be retrieved for use in later SAS sessions.

variable

a set of data values in a SAS data set that describe a given characteristic.

warning

in SAS logs, a message that informs you of a potential problem.

Index

Your Turn

If you have comments or suggestions about *SAS Language and Procedures: Introduction, Version 6, First Edition*, please send them to us on a photocopy of this page.

Please return the photocopy to the Publications Division (for comments about this book) or the Technical Support Division (for suggestions about the software) at SAS Institute Inc., SAS Campus Drive, Cary, NC 27513.

Experience is the Best Teacher

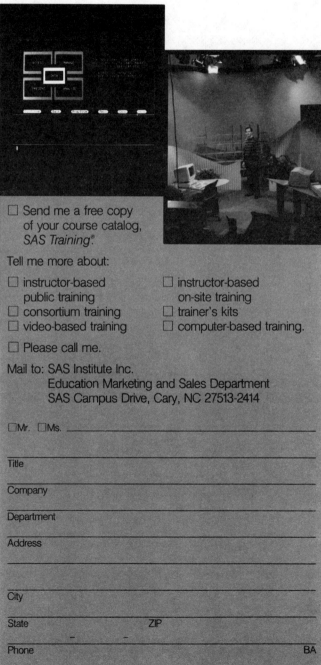